CW00733167

The Institute of Mathematics
and its Applications
Conference Series

Volumes in the previous series were published by Academic Press to whom all enquiries should be addressed. The following and all forthcoming titles are published by Oxford University Press throughout the world.

NEW SERIES

1. *Supercomputers and parallel computation* Edited by D. J. Paddon
2. *The mathematical basis of finite element methods* Edited by David F. Griffiths
3. *Multigrid methods for integral and differential equations* Edited by D. J. Paddon and H. Holstein
4. *Turbulence and diffusion in stable environments* Edited by J. C. R. Hunt
5. *Wave propagation and scattering* Edited by B. J. Uscinski
6. *The mathematics of surfaces* Edited by J. A. Gregory
7. *Numerical methods for fluid dynamics II* Edited by K. W. Morton and M. J. Baines
8. *Analysing conflict and its resolution* Edited by P. G. Bennett
9. *The state of the art in numerical analysis* Edited by A. Iserles and M. J. D. Powell
10. *Algorithms for approximation* Edited by J. C. Mason and M. G. Cox
11. *The mathematics of surfaces II* Edited by R. R. Martin
12. *Mathematics in signal processing* Edited by T. S. Durrani, J. B. Abbiss, J. E. Hudson, R. N. Madan, J. G. McWhirter, and T. A. Moore
13. *Simulation and optimization of large systems* Edited by Andrzej J. Osiadacz
14. *Computers in mathematical research* Edited by N. M. Stephens and M. P. Thorne
15. *Stably stratified flow and dense gas dispersion* Edited by J. S. Puttock
16. *Mathematical modelling in non-destructive testing* Edited by Michael Blakemore and George A. Georgiou
17. *Numerical methods for fluid dynamics III* Edited by K. W. Morton and M. J. Baines
18. *Mathematics in oil production* Edited by Sir Sam Edwards and P. R. King
19. *Mathematics in major accident risk assessment* Edited by R. A. Cox
20. *Cryptography and coding* Edited by Henry J. Beker and F. C. Piper
21. *Mathematics in remote sensing* Edited by S. R. Brooks
22. *Applications of matrix theory* Edited by M. J. C. Gover and S. Barnett
23. *The mathematics of surfaces III* Edited by D. C. Handscomb
24. *The interface of mathematics and particle physics* Edited by D. G. Quillen, G. B. Segal, and Tsou S. T.
25. *Computational methods in aeronautical fluid dynamics* Edited by P. Stow
26. *Mathematics in signal processing II* Edited by J. G. McWhirter
27. *Mathematical structures for software engineering* Edited by Bernard de Neumann, Dan Simpson, and Gil Slater
28. *Computer modelling in the environmental sciences* Edited by D. G. Farmer and M. J. Rycroft
29. *Statistics in medicine* Edited by F. Dunstan and J. Pickles
30. *The mathematical revolution inspired by computing* Edited by J. H. Johnson and M. J. Loomes
31. *The mathematics of oil recovery* Edited by P. R. King
32. *Credit scoring and credit control* Edited by L. C. Thomas, J. N. Crook, and D. B. Edelman
33. *Cryptography and coding II* Edited by Chris Mitchell
34. *The dynamics of numerics and numerics of dynamics* Edited by D. S. Broomhead and A. Iserles
35. *The unified computation laboratory: modelling, specifications, and tools* Edited by Charles Rattray and Robert G. Clark

Continued overleaf

36. *Mathematics in the automotive industry* Edited by James R. Smith
37. *The mathematics of control theory* Edited by N. K. Nichols and D. H. Owens
38. *Mathematics in transport planning and control* Edited by J. D. Griffiths
39. *Computational ordinary differential equations* Edited by J. R. Cash and I. Gladwell
40. *Waves and turbulence in stably stratified flows* Edited by S. D. Mobbs and J. C. King
41. *Robotics: applied mathematics and computational aspects* Edited by Kevin Warwick
42. *Mathematical modelling for materials processing*
    Edited by Mark Cross, J. F. T. Pittman, and Richard D. Wood
43. *Wavelets, fractals, and Fourier transforms*
    Edited by M. Farge, J. C. R. Hunt, and J. C. Vassilicos
44. *Computational aeronautical fluid dynamics* Edited by J. Fezoui, J. Hunt, and J. Periaux
45. *Cryptography and coding III* Edited by M. J. Ganley
46. *Parallel computation* Edited by A. E. Fincham and B. Ford
47. *Aerospace vehicle dynamics* Edited by M. V. Cook and M. J. Rycroft
48. *Computer-aided surface geometry and design: mathematics of surfaces IV*
    Edited by A. Bowyer
49. *Mathematics in signal processing III* Edited by J. G. McWhirter
50. *Design and application of curves and surfaces: mathematics of surfaces V*
    Edited by R. B. Fisher
51. *Artificial intelligence in mathematics* Edited by J. H. Johnson, S. McKee, and A. Vella
52. *Stably stratified flows: flow and dispersion over topography*
    Edited by I. P. Castro and N. Rockliffe
53. *Computational learning theory: EuroCOLT 93* Edited by J. S. Shawe-Taylor and M. Antony
54. *Complex stochastic systems and engineering* Edited by D. M. Titterington
55. *Mathematics of dependable systems* Edited by C. J. Mitchell and V. Stavridou
56. *The mathematics of deforming surfaces* Edited by D. G. Dritschel and R. J. Perkins
57. *Mathematical education of engineers* Edited by L. R. Mustoe and S. Hibberd
58. *The mathematics of surfaces VI* Edited by G. Mullineux
59. *Applications of finite fields* Edited by Dieter Gollmann
60. *Applications of combinatorial mathematics* Edited by C. J. Mitchell
61. *Image processing: mathematical methods and applications* Edited by J. M. Blackledge
62. *Flow and dispersion through groups of obstacles* Edited by R. J. Perkins and S. E. Belcher
63. *The state of the art in numerical analysis* Edited by I. S. Duff and G. A. Watson
64. *Mathematics of dependable systems II* Edited by V. Stavridou
65. *Approximations and numerical methods for the solution of Maxwell's equations*
    Edited by F. El Dabaghi, K. Morgan, A. K. Parrott, and J. Periaux
66. *Mathematics of heat transfer* Edited by G. E. Tupholme and A. S. Wood
67. *Mathematics in signal processing IV* Edited by J. G. McWhirter and I. K. Proudler
68. *Mixing and dispersion in stably stratified flows* Edited by Peter A. Davies
69. *Wind-over-wave couplings: perspectives and prospects* Edited by S. G. Sajjadi, N. H. Thomas, and J. C. R. Hunt
70. *Cardiovascular flow modelling and measurement with application to clinical medicine*
    Edited by S. G. Sajjadi, G. B. Nash, and M. W. Rampling

Cardiovascular Flow Modelling and Measurement with Application to Clinical Medicine

Based on the proceedings of a conference organized by the Institute of Mathematics and its Applications on Cardiovascular Flow Modelling and Measurement with Application to Clinical Medicine, and held at the University of Salford in September 1998.

Edited by

S. G. SAJJADI

Centre for Computational Fluid Dynamics and Turbulence
University of Salford

G. B. NASH

Department of Physiology, The Medical School
The University of Birmingham

M. W. RAMPLING

Imperial College School of Medicine

CLARENDON PRESS • OXFORD • 1999

OXFORD
UNIVERSITY PRESS

Great Clarendon Street, Oxford OX2 6DP

Oxford University Press is a department of the University of Oxford.
It furthers the University's objective of excellence in research, scholarship,
and education by publishing worldwide in

Oxford New York

Athens Auckland Bangkok Bogotá Buenos Aires Calcutta
Cape Town Chennai Dar es Salaam Delhi Florence Hong Kong Istanbul
Karachi Kuala Lumpur Madrid Melbourne Mexico City Mumbai
Nairobi Paris São Paulo Singapore Taipei Tokyo Toronto Warsaw

with associated companies in Berlin Ibadan

Published in the United States
by Oxford University Press Inc., New York

First published 1999

A catalogue record for this book is available from the British Library

Library of Congress Cataloging in Publication Data

(Data applied for)

ISBN 0 19 850520 5

Typeset using LaTeX
Printed in Great Britain
on acid-free paper by
Bookcraft (Bath) Ltd., Midsomer Norton, Avon

FOREWORD

Trends in Cardiovascular Fluid Dynamics
by T.J. Pedley, FRS, C.Math. FIMA

In the 1950s and 1960s the principal research topic in arterial fluid dynamics was the propagation and reflection of the pressure pulse, and the corresponding flow-rate or velocity pulse. It was hoped that an understanding of vascular mechanics, in particular the factors determining the effective, or input, impedance of the aorta or another major artery would lead to improved diagnosis of systemic vascular disease. This phase of endeavour was epitomised by McDonald's classic book, *Blood Flow in Arteries* (McDonald, 1960), which grew out of his research collaboration with the physicist, J.R. Womersley.

Thirty years ago, however, the focus of attention switched to measurement and analysis of the detailed flow patterns and distributions of wall shear stress (WSS) within arteries. This came about as a result of two seminal contributions: by D.L. Fry (1968), who noted that arterial endothelial cells could be damaged by severely *elevated* levels of WSS; and by C.G. Caro and colleagues (Caro, et al., 1971) who observed a rough correlation between arterial sites which are prone to the development of arterial disease (atherosclerosis) and sites at which the mean level of WSS would be expected to be *low*. This expectation was based on a rather crude view of arterial fluid dynamics, but it has stood the test of time, and the low-mean-shear hypothesis for the initiation of atherosclerosis has largely prevailed over the high-shear hypothesis, although there is evidence that time-dependence of the flow (in the form of repeated flow reversals) also has an effect (Ku, et al., 1985).

The low-shear/high-shear debate was a tremendous stimulus to research in arterial fluid mechanics, as well as in vascular wall biology. The aims of the latter have been to reveal the biological and structural effects of various levels of WSS on endothelial cells (in vitro as well as in vivo) and to try to understand how they come about (Davies, 1995). The aims of the former are to reveal and understand the actual distribution of WSS in the complex three-dimensional geometry of large arteries. These aims have spawned a huge number of papers in the fluid dynamics and bioengineering literature, describing model experiments, numerical computations and mathematical analysis of the flow in a sequence of increasingly complex geometries representative of normal or diseased arteries and of surgical modifications to them (such as femoral bypass grafts): entry flow in straight tubes; flow in constricted tubes; fully-developed and entry flow in uniform curved tubes; flow in helical tubes; flow in moving tubes; flow in symmetric planar bifurcations; flow in all of the above when the vessels are elastic instead of rigid; using non-Newtonian as well as Newtonian models of blood rheology; investigating unsteady as well as steady flow. Since the mean Reynolds number in large arteries is quite large (several hundred, typically) the flow and WSS distribution are extremely sensitive to small geometrical perturbations and to small changes in the time-course, as represented by the wave-form of pressure-gradient or flow-rate. Steady flow can be non-unique, and stable or unstable, though rarely fully turbulent. See Pedley (1995) for a review and further references.

Such sensitivity to the precise spatial or temporal details raises doubts about the predictive value of many of the numerical or experimental simulations that are performed on (for example) home-made bifurcation models, or even casts of an individual subjects' vessels. They reveal great complexity of flow, which in itself is instinctive, but since one subject's complexity is likely to be different from another's, what is to be gained (other than technical expertise) from the quantitative minutiae? In my view really deep quantitative investigations should be restricted to two extremes of the spectrum.

The first is the applied mathematician's approach: examine in great thoroughness highly idealised geometries which can be readily reproduced by other investigators and in each of which a single physical mechanism or phenomenon is exemplified. An excellent example is steady or pulsatile flow in a uniform curved tube with circular cross-section. All sorts of fluid dynamical features can be found here: the appearance of secondary (transverse) velocities when the driving pressure gradient is purely longitudinal; the development of boundary-layer singularities in a flow started from rest; secondary steady streaming in a flow driven by a purely sinusoidal pressure gradient; non-uniqueness of steady flow at values of the Reynolds number which are not excessively large; inhibition of transition to turbulence; reduction of the entry length (relative to that in a straight tube), etc. Many of the phenomena are still not fully understood because the subtle structure of the flow defeats any simple-minded attempt to simulate it numerically. There is still much to be learnt from such problems.

The other approach, currently being followed by several groups in North America and elsewhere, is to seek to simulate the flow in each individual patient and predict the WSS distribution for use in diagnosis or surgical planning (it goes without saying that direct and accurate measurement of WSS is not feasible using any known technique). This requires direct imaging of the geometry, for example using magnetic resonance imaging (MRI). Then the MR data, "suitably" smoothed (a process about which there is considerable controversy), must be automatically converted to a computational mesh, on which CFD analysis will be performed (usually using finite-element or finite-volume methods). Input to the calculations in the form of two-point pressure or, more usually, one-point flow-rate wave forms also needs to be measured. Again, MR methods are increasingly being used for flow rate measurement, though ultra-sound is currently more reliable. At present every stage of this process has problems and raises doubts about its validity and accuracy. Nevertheless, the development of such integrated approaches is an area of great current excitement, and I am sure that within about five years we will see reliable data emerging routinely for clinical use with individual patients: but it will be expensive in computer resources. See ASME (1999), especially pages 717–728, for some very recent abstracts in this area.

I have dwelt on arterial fluid mechanics because that is the aspect of Cardiovascular Flow Modelling that dominates the international biomechanics community and is the most liberally funded because of the relevance to vascular disease. It is also the area I know most about. However, we should not forget the venous system, which contains up to 80% of the blood volume, albeit at relatively low pressure. It is less intensely studied because it is not normally associated with life-threatening diseases apart from the thrombus formation and subsequent embolism that may occur when a patient remains immobile for too long. However, the venous system has some interesting features, not shared by arteries. These include non-return valves in dependent limbs, which have received rather little attention, and vessel collapse above the level of the heart, where gravity causes the internal pressure to fall below the external pressure (a phenomenon that is particularly marked in the giraffe; see Pedley, et al., 1996).

The microcirculation, too, is the subject of extensive biomechanical research. The main complication, from the fluid dynamics point of view, is that blood is a concentrated ($\sim$ 45% by volume) suspension of cells in a virtually Newtonian fluid, the plasma. This means both that whole blood does not behave as a Newtonian fluid (it is shear-thinning for shear-rates less than about $100s^{1}$), and that it cannot be regarded as a homogeneous medium at length scales comparable with those of the cells. In particular, red and white blood cells (diameter $\sim 8\mu$m) must enter narrow capillaries one at a time and are considerably deformed as they travel down them in single file (see Caro, et al., 1978). Variations in the rheological properties of blood turn out to be sensitive indicators of a range of diseases, and the measurement and interpretation of such properties forms the basis of the field of (Clinical) Haemorheology.

No part of cardiovascular fluid mechanics can be studied in isolation by a theoretician. The field is interdisciplinary and necessarily involves strong collaboration between engineers or physical scientists (including applied mathematicians) on the one hand and physicians or biological scientists on the other. The partnership between McDonald and Womersley was an excellent example of such intimate collaboration.

The Physiological Flow Studies Unit, formed at Imperial College, London in 1966 under the directorship of C.G. Caro but with the help of Sir James Lighthill, was another; it now forms a part of the Department of Biological and Medical Systems there. The IMA Conference on Cardiovascular Flow Modelling and Measurement with Application to Clinical Medicine, of which this volume constitutes the proceedings, also represented a real effort to bring the engineers (mostly working on arterial flow) and clinicians (mostly involved in haemorheology) together, and was extremely successful in that aim. Both groups learned a lot from each other. The biggest group of papers in this book is on arterial flow, with first authors (in alphabetical order): Bates, Cole, David, Doorly, Gerrard (the only one to concentrate exclusively on pulse wave propagation), Kirkham, Migliavacca, Neofitou, Sajjadi, Spentzos and Zhao, most but not all from engineering departments or their equivalent. The haemorheology group of papers is smaller (Evans, Kesmarky, Nash, Rampling and Toth) and more clinically oriented but equally fascinating. It is to be hoped that this exercise in collaboration between the different groups helps to stimulate new joint activity, both between those who were at the conference and among those who learn about it first from these pages.

References and Further Reading

ASME (1999). *Proceedings of the 1999 Bioengineering Conference*, Editors: V.K. Goel, R.L. Spilker, G.A. Ateshian, and L.J. Soslowsky, **BED–Volume 42**, American Society of Mechanical Engineers.

Caro, C.G., Fitz-Gerald, J.M., and Schroter, R.C. (1971). Atheroma and arterial wall shear: observation, correlation and proposal of a shear dependent mass transfer mechanism for atherogenesis. *Proc. R. Soc. Lond.*, **B177**, pp. 109–159.

Caro, C.G., Pedley, T.J., Schroter, R.C., and Seed, W.A. (1978). *The Mechanics of the Circulation*, Oxford University Press.

Fry, D.L. (1968). Acute vascular endothelial changes associated with increased blood velocity gradients. *Circ. Res.*, **22**, pp. 165–197.

Ku, D.N., Giddens, D.P., Zarins, C.K., and Glagov, S. (1985). Pulsatile flow and atherosclerosis in the human carotid bifurcation: positive correlation between plaque location and low and oscillating shear stress. *Arteriosclerosis*, **5**, pp. 293–302.

McDonald, D.A. (1960). *Blood Flow in Arteries*, (and 1974, 2nd Edition), Edward Arnold, London.

Pedley, T.J. (1980). *The Fluid Mechanics of Large Blood Vessels*, Cambridge University Press.

Pedley, T.J. (1995). High Reynolds number flow in tubes of complex geometry with application to wall shear stress in arteries. *Biological Fluid Dynamics*, Editors: C.P. Ellington and T.J. Pedley, **SEB Symposium No. 49**, The Company of Biologists.

Pedley, T.J., Brook, B.S., and Seymour, R.S. (1996). Blood pressure and flow rate in the giraffe jugular vein. *Phil. Trans. R. Soc. Lond.*, **B351**, pp. 855–866.

PREFACE

Interdisciplinary scientific research has played a significant role in the advancement of knowledge, especially in the biosciences. Such interdisciplinary research has arisen from the necessity to investigate phenomena in one branch of science by linking it to another discipline.

The cardiovascular system has been the subject of extensive studies by the medical profession and over the years a vast amount of knowledge has been accumulated. However, during recent years, there has been a tremendous attempt to try to study and model the cardiovascular system from the mathematical, fluid mechanical and rheological point of view.

The papers here were presented at a conference held in September 1998 at the University of Salford organised by the Institute of Mathematics and its Applications (IMA) with the co-operation of members of the Committee of the International Society of Clinical Haemorheology. The aim of this conference was to simulate and coordinate research in this rapidly developing subject and especially to bring together the many different groups of specialists who are now involved.

This book is a collection of what we believe to be useful papers on the main research topics of cardiovascular modelling and measurements, that are current at the end of the 1990s. Some of the results presented here show that we can now calculate many non-linear aspects of fluid flow, and most significantly, the turbulence in the arteries, bifurcation junctions of the cardiovascular system. Of course, as with experimental results the insights derived from these calculations have given rise to deeper understanding and the scope for further research. Research into the cardiovascular system is also being transformed by measurements of haemorheological factors. However, as we have access to ever more computational and measurement capability, new questions arise, about the details of the flow in arteries, and even about the nature of the complex interaction that takes place in the blood. These questions are steadily changing the nature of the subject of haemodynamics and haemorheology.

There are three main themes to the papers included in this volume. The first deals with the fundamental concepts, namely the fluid dynamics and turbulence in cardiovascular system. The paper here by Gerrard reviews the standard Womersley theory for pulsatile flow in destensible tube with extension to walls which are viscoelastic and not infintesimally thin. He showed that there are two wave types and discussed the possibility of each wave type producing the other. Other papers focus on flow stability and turbulence in suddenly-expanded channels and tubes. The Direct Numerical Simulations described by Spentzos and Drikakis revealed that for Reynolds numbers in the stability regime, an initial random flow perturbation is quickly attenuated. The theoretical calculations of Sajjadi and Feng examined pulsatile flow in an elliptical tube with stenotic constriction. They showed that the characteristics of the flow are affected by the percentage change in stenosis. They also confirmed their analytical findings with a numerical model of turbulent flow over stenosis. The effect of unsteady flow in a channel with a laterally moving indentation was examined by Neofitou, Drikakis and Leschziner. In this paper comparisons were made between Newtonian and non-Newtonian flow which reveal differences in flow structure, especially regarding the formation of the vorticity waves downstream of the indentation. Two further papers examined the effect of platelet deposition in stagnation point flow, and fluid dynamics of cavopulmonary connections with extracardiac lateral conduit. The simulation by David, Thomas and Walker showed that for a constant wall reaction rate modelling platelet adhesion the maximum platelet attachment occurs around the stagnation point streamline. Migliavacco et al. reported three-dimensional fluid dynamics models of cavopulmonary connections. Their results showed that left-to-right pulmonary flow ratio and percentage inferior caval blood to the left lung were the highest with the smallest anastomosis.

The second theme is the flow modelling in arteries and bypass graft. Doorly here reviewed the modelling of flow transport in large arteries using hybrid Eulerian-Lagrangian techniques and argued that within large arteries, the dominant means of transport is by convection. He also outlined the extension of his computational procedure to enable simulation of magnetic resonance images of flow in arteries. The application of computational fluid dynamics to haemodynamics in arterial organs was the subject of a lecture by Collins. In this paper he discussed a range of topics for correlating haemodynamics to risk factors which underlines the genesis and progression of cardiovascular disease. In the paper by Bates, Williams and O'Doherty an experimental and computational modelling of flow through an arterial bypass graft was presented. They argued that fluctuating wall shear stresses in the graft are believed to be important factors in the development and localisation of intimal hyperplasia. Zhao et al. presented a technique in numerical solutions of coupled fluid-wall problems. In this paper they discussed some important issues related to the coupled model, particularly numerical convergence. Modelling of cerebral autoregulation experiments in humans was the subject of the lecture by Kirkham. In this paper she showed how the knowledge of a patient's ability to autoregulate could lead to an improved understanding of their condition and thereby help to manage their therapy. The final paper in this group by Cole et al. focused on the numerical investigation of a novel haemodynamics control device which reduces the development of occlusive arterial intimal hyperplasia.

For the final group of papers we have five on haemorheology and heamodynamics. In the paper by Toth et al. the haemorheological factors that play an important role in the coronary circulation were reviewed. Using their own data, on close to 1000 patients with ischemic heart disease (IHD) and acute myocardial infection (AMI), they showed that the rheological parameters in IHD and AMI are in the pathological range and are significantly higher than in healthy controls. Haemorheological alterations after percutaneous transluminal coronary angioplasty (PTCA) was the subject of the paper by Kesmarky et al. Their findings indicated that PTCA may cause significant changes in the haemorheological parameters which can affect the final result of intervention. Nash et al. discussed adhesive interaction between leukocytes and platelets at the vessal wall. Using in vitro flow models they suggested that co-deposition can occur in arteries, and an understanding of how this comes about could significantly contribute to our understanding of atherosclerosis. The causes and consequences of rouleaux formation were reviewed by Rampling. From a rheological point of view he argued that the rouleaux formation is a major factor responsible for the remarkable shear dependence of blood viscosity. In the final paper of this group, Evans and Cook investigated monocyte rheology in which blood obtained from two groups of healthy volunteers was filtered using a fully automated constant pressure filtrometer. The flow profiles were then analysed by least squares technique to an appropriate mathematical model, to predict the contribution of these cells to microvascular flow resistance.

The editors would particularly like to thank the other members of the International Organising Committee for suggesting invited speakers and topics, and the invited speakers for their contributions. We would particularly like to thank Professor Tim Pedley, FRS, who kindly agreed to write the Foreword for this volume. We would also like to thank all the speakers for their efforts in preparing and delivering their talks, and we are grateful to those who so ably assisted as Chairman for various sessions. We are grateful to Pamela Bye, of the IMA, for her help in the organisation. The conference was co-sponsored by the International Society for Clinical Haemorheology, the International Society for Biorheology and the Royal Society of Medicine Forum on Angiology, and we gratefully acknowledge their assistance.

For the production of this volume, we would like to thank all those who contributed, for their co-operation, for their patience, and their willingness to cope with the technicalities of achieving uniformity of style. The editors sought comments on all the papers from appropriate experts in the field, and we are grateful to them, and also to the authors, for responding so readily to suggestions which were relayed to them. The papers were checked and put into final camera ready copy, at the IMA, by Debbie Brown before their final journey to Oxford University Press. We would like to express our thanks to her and the individuals concerned at Oxford University Press.

Shahrdad G. Sajjadi
University of Salford

Gerard B. Nash
University of Birmingham

Michael W. Rampling
Imperial College, London

CONTENTS

CONTRIBUTORS

C.J. BATES; Division of Mechanical Engineering and Energy Studies, Cardiff University, Queens Building, The Parade, P.O. Box 685, Cardiff, Wales, CF2 3TA.

A.A. BIRCH; Department of Medical Physics and Bioengineering, Southampton University Hospitals NHS Trust, Southampton.

J.S. COLE; School of Aeronautical Engineering, Queen's University of Belfast, David Keir Building, Stranmillis Road, Belfast, Northern Ireland, BT9 5AG.

M.W. COLLINS; School of Engineering Systems and Design, South Bank University, London.

A. COOK; Cardiff School of Biosciences, Cardiff University, Museum Avenue, P.O. Box 911, Cardiff, Wales, CF1 3US.

R.E. CRAINE; Faculty of Mathematical Studies, University of Southampton, Highfield, Southampton, SO17 1BJ.

T. DAVID; School of Mechanical Engineering, University of Leeds, Leeds, LS2 9JT.

N.R. DE LEVAL; Cardiothoracic Unit, Great Ormond Street Hospital for Children, Great Ormond Street, London, WC1N 3JH.

D.J. DOORLY; Department of Aeronautics, Imperial College, Prince Consort Road, London, SW7 2BY.

D. DRIKAKIS; Department of Engineering, Queen Mary and Westfield College, University of London, London, E1 4NS.

G. DUBINI; Department of Energetics, Politecnico di Milano, Piazza Leonardo da Vinci 32, 20133 Milano, Italy.

S-A. EVANS; Cardiff School of Biosciences, Cardiff University, Museum Avenue, P.O. Box 911, Cardiff, Wales, CF1 3US.

Y. FENG; Centre for Computational Fluid Dynamics and Turbulence, University of Salford, Salford, M5 4WT.

R. FUMERO; Department of Bioengineering, Politecnico di Milano, Piazza Leonardo da Vinci 32, 20133 Milano, Italy.

J.H. GERRARD; Division of Aerospace Engineering, Manchester School of Engineering, University of Manchester, Goldstein Laboratory, Barton Airport, Manchester, M30 7RU.

M.A. GILLAN; School of Aeronautical Engineering, Queen's University of Belfast, David Keir Building, Stranmillis Road, Belfast, Northern Ireland, BT9 5AG.

L. HABON; Department of Surgery, Heart Center, University Medical School of Pecs, Ifjusag ut 13, H-7643 Pecs, Hungary.

T. HABON; Division of Cardiology, Department of Medicine, University Medical School of Pecs, Ifjusag ut 13, H-7643 Pecs, Hungary.

R. HALMOSI; Division of Cardiology, Department of Medicine, University Medical School of Pecs, Ifjusag ut 13, H-7643 Pecs, Hungary.

A.D. HUGHES; Department of Clinical Pharmacology, Imperial College School of Medicine at St. Mary's, London.

I. JURICSKAY; Division of Cardiology, Department of Medicine, University Medical School of Pecs, Ifjusag ut 13, H-7643 Pecs, Hungary.

G. KESMARKY; Division of Cardiology, Department of Medicine, University Medical School of Pecs, Ifjusag ut 13, H-7643 Pecs, Hungary.

S.K. KIRKHAM; Faculty of Mathematical Studies, University of Southampton, Highfield, Southampton, SO17 1BJ.

C.M. KIRTON; Department of Physiology, The Medical School, University of Birmingham, Birmingham, B15 2TT.

M.A. LESCHZINER; Department of Mechanical Engineering, University of Manchester Institute of Science and Technology, P.O. Box 88, Manchester, M60 1QD.

Q. LONG; Department of Chemical Engineering and Chemical Technology, Imperial College, Prince Consort Road, London, SW7 2BY.

A. McKINLEY; Vascular Unit, Belfast City Hospital, Belfast, Northern Ireland, BT9 7AB.

F. MIGLIAVACCA; Department of Bioengineering, Politecnico di Milano, Piazza Leonardo da Vinci 32, 20133 Milano, Italy.

G.B. NASH; Department of Physiology, The Medical School, University of Birmingham, Birmingham, B15 2TT.

P. NEOFITOU; Department of Mechanical Engineering, University of Manchester Institute of Science and Technology, P.O. Box 88, Manchester, M60 1QD.

D.M. O'DOHERTY; Division of Mechanical Engineering and Energy Studies, Cardiff University, Queens Building, The Parade, P.O. Box 685, Cardiff, Wales, CF2 3TA.

M.J.G. O'REILLY; Vascular Unit, Belfast City Hospital, Belfast, Northern Ireland, BT9 7AB.

R. PIETRABISSA; Department of Bioengineering, Politecnico di Milano, Piazza Leonardo da Vinci 32, 20133 Milano, Italy.

S. RAGHUNATHAN; School of Aeronautical Engineering, Queen's University of Belfast, David Keir Building, Stranmillis Road, Belfast, Northern Ireland, BT9 5AG.

G.E. RAINGER; Department of Physiology, The Medical School, University of Birmingham, Birmingham, B15 2TT.

M.W. RAMPLING; Imperial College School of Medicine, Sir Alexander Fleming Building, South Kensington, London, SW7 2AZ.

E. ROTH; Department of Experimental Surgery, University Medicical School of Pecs, Kodaly z. ut 20, H-7624, Hungary.

S.G. SAJJADI; Centre for Computational Fluid Dynamics and Turbulence, University of Salford, Salford, M5 4WT.

A. SPENTZOS; Department of Mechanical Engineering, University of Manchester Institute of Science and Technology, P.O. Box 88, Manchester, M60 1QD.

A.V. STANTON; Department of Clinical Pharmacology, Imperial College School of Medicine at St. Mary's, London.

P.C.W. STONE; Department of Physiology, The Medical School, University of Birmingham, Birmingham, B15 2TT.

S.A. THOM; Department of Clinical Pharmacology, Imperial College School of Medicine at St. Mary's, London.

S. THOMAS; School of Mechanical Engineering, University of Leeds, Leeds, LS2 9JT.

K. TOTH; Division of Cardiology, Department of Medicine, University Medical School of Pecs, Ifjusag ut 13, H-7643 Pecs, Hungary.

G. VAJDA; Department of Surgery, Heart Center, University Medical School of Pecs, Ifjusag ut 13, H-7643 Pecs, Hungary.

P.G. WALKER; School of Mechanical Engineering, University of Leeds, Leeds, LS2 9JT.

D. WILLIAMS; Division of Mechanical Engineering and Energy Studies, Cardiff University, Queens Building, The Parade, P.O. Box 685, Cardiff, Wales, CF2 3TA.

X.Y. Xu; Department of Chemical Engineering and Chemical Technology, Imperial College, Prince Consort Road, London, SW7 2BY.

S. Z. ZHAO; Department of Chemical Engineering and Chemical Technology, Imperial College, Prince Consort Road, London, SW7 2BY.

Boundary Conditions and the Attenuation of Waves in Deformable Tubes

J.H. Gerrard

Manchester School of Engineering, Goldstein Laboratory, Manchester University

Abstract

The standard Womersley theory for pulsatile flow in distensible tubes is outlined with entension to walls which are not infinitesimally thin and which are viscoelastic. In general there are two waves types, one in which the wall motion is principally radial (waves 1) and those in which the motion is principally longitudinal (waves 2). When the oscillating motion is produced by a piston driving one end of the tube it is found that waves 1 are first produced and these produce waves 2. The possibility of each wave type producing the other is discussed. Waves 2 appear not to be found in cardiovasculcar systems.

The paper concludes with the analysis of waves in a tethered tube and the boundary conditions are discussed.

1 Introduction

My subject is fluid mechanics. In 1965 I became interested in cardiovascular flow having heard some lectures on the subject. There used to be a Medical Engineering Club in Manchester at which liaison between scientists and clinicians was encouraged. I did a little work with Derek Rowlands at the Manchester Royal Infirmary and later Professor David Charlesworth and I got together at the University Hospital of South Manchester in Withington. He had a well equipped laboratory in which we looked at pulsatile flow in latex tubes. This has been followed by many years of collaboration. Our latest interest has been in atherogenesis. At this meeting it does not need to be stressed that collaboration between scientists and clinicians is a vital constituent for progress.

Theoretical work on flow in tubes has been spread over many years following the initial work of Young in 1808 and 1809. Perhaps surprisingly Young was a clinician. We can list the primary workers in the field with dates as follows: Korteweg (1878), Lamb (1897), Witzig (1914), King (1947), Lambossy (1950), Müller (1951), Love (1952), Morgan and Kiely (1954), Morgan and Ferrante (1955), Womersley (1957) and Taylor (1959). The starting point of most modern work is that of Womersley which is essentially a repeat of Witzig's work which lay undiscovered in a library in Berne for many years. Whether Womersley knew of it I do not know.

2 Womersley theory

We consider a thin walled tube of radius R shown in Figure 1 from Gerrard [1]. All variables are of the form

$$p = \hat{p}\exp(in(t - z/c))$$

where c is the complex wave speed or alternatively in terms of the propagtion constant γ

$$p = \hat{p}\exp(int - \gamma z)$$

$$\gamma = {}^{in}/c = \gamma_r + {}^{in}/w$$

where γ_r is the real part and w is the real wave speed. Small amplitude is considered and the wave length $\lambda >> R$ the tube radius. The fluid equations are linear and velocities are obtained in terms of $\hat{p}$ and a constant C_1.
The wall equations are

$$\begin{aligned} m\ddot{\zeta} &= \text{elastic forces + skin fraction force of the fluid} \\ &\quad \text{at } y = 1 \text{ and a suspension restraint} \\ m\ddot{\xi} &= \text{elastic forces + force of pressure in the tube.} \end{aligned}$$

When the tube moves longitudinally there should be a friction force on the outside as well as the inside of the tube. Matching at $y = 1$ gives $w = \dot{\zeta}$ and $u = \dot{\xi}$. Four equations are obtained in ${}^{C_1}/_{\hat{p}}$, ${}^{\xi}/_{\hat{p}}$, ${}^{\zeta}/_{\hat{p}}$ and c or γ.

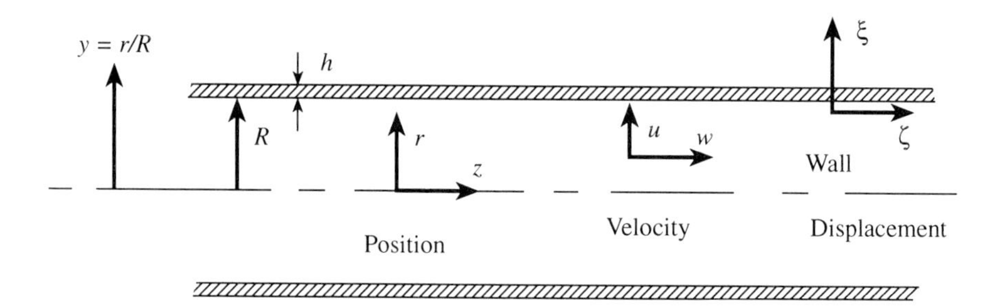

Figure 1. The variables of the tube and motion

Table 1. Expression for variables in terms of pressure amplitude p_1

In Womersley's variables	For Poisson's ratio = 0.5	In terms of γ
$C_1/(p_1/\rho c^2) = \eta = \frac{2\sigma-1+2/x}{F_{10}-2\sigma}$	$= \frac{-2}{x(1-F_{10})}$	$= \frac{3n^2}{4\gamma^2 c_0^2(1-F)_{10})}$
$\frac{\xi_1}{R} = \frac{p_1}{\rho c^2}\frac{\sigma(F_{10}-1+F_{10}/\sigma x)}{F_{10}-2\sigma}$	$= \frac{p_1}{2\rho c^2}\left\{1-\frac{2F_{10}}{x(1-F_{10})}\right\}$	$= \frac{-p_1\gamma^2(1+\eta F_{10})}{2\rho n^2}$
$\xi_1 = \hat{\xi}$		
$\frac{n\zeta_1}{c} = \frac{p_1}{\rho c^2}\frac{i(1-F_{10}-2/x)}{F_{10}-2\sigma}$	$\zeta_1 = \frac{-ip_1(1+\eta)}{\rho n c}$	$= \frac{-\gamma p_1(1+\eta)}{\rho n^2}$
$\overline{w}_1 = \frac{p_1}{\rho c}(1+\eta F_{10})$		$= \frac{-i\gamma p_1(1+\eta F_{10})}{\rho n}$

For a tethered tube ($\zeta = 0$)

$$\eta = -1,\ p_1 = \frac{2\rho c_0^2}{1-\sigma^2}\xi_1/R$$

$\overline{w}_1$ is the amplitude of the cross-sectional mean speed of the fluid; ρ is the fluid density which Womersley calls ρ_o.
$F_{10} = 2J_1(z)/(zJ_0(z))$ where $z = \alpha i^{3/2}$, J_1 and J_0 are Bessel functions of the first kind of order 1 and 0.

The form of the solution is $\zeta = \hat{\zeta}_1 e^{\text{int}\pm\gamma_1 z} + \hat{\zeta}_2 e^{\text{int}\pm\gamma_2 z}$ and similarly for ξ. Suffices 1 and 2 apply to the two types of wave: wave 1 in which the motion is principally radial and wave 2 in which it is principally longitudinal.
Womersley's solutions are in terms of quantities which he calls c, c_0, and x where

$$c_0 \text{ is the Moens - Korteweg speed } = \left(\frac{Eh}{2\rho R}\right)^{1_2}.$$

h is the wall thickness and E Young's modulus. The Womersley parameter, $\alpha = R\sqrt{n/\nu}$. For an inviscid fluid α is infinite.
Table 1 from Gerrard [1] gives expressions in terms of x which Womersley chooses to use.

$$x = \frac{2}{1-\sigma^2}\frac{c_o^2}{c^2} = \frac{-2}{1-\sigma^2}\frac{\gamma^2 c_o^2}{n^2} \tag{2.1}$$

where σ is Poisson's ratio which for latex rubber is 0.5. Rubber is incomparable (but easily deformable). We consider now a thick walled tube. Static experiments on a latex tube show that

$$p = \frac{2}{3}\rho c_0^2 \vartheta_o (1 - D_0^4/D^4) \tag{2.2}$$

D_o and D are the inside and outside diameter of the tube. Taylor and Gerrard [3] showed that c_o is replaced by $c_0\sqrt{\vartheta_o}$ where

$$\vartheta_0 = \left(1 + ^{h_o}/D_o\right) / \left(1 + \frac{2h_o}{D_o}\right)^2. \tag{2.3}$$

For $h_o = 0.18$ cm, $D_o = 0.63$ cm, $\vartheta_0 = 0.52$. The linearised form of Equation (2.2) is

$$p = \frac{8}{3}\rho c_o^2 \vartheta_0 \left(\frac{D}{D_o} - 1\right). \tag{2.4}$$

With ϑ_o and viscoelasticity included agreement with experiment in a tethered tube is achieved, as we shall see. The equation quoted by Wormersley is

$$p = \frac{2\rho c_o^2 (^D/_{Do} - 1)}{c_o^2/c^2 - 0.75\frac{F_{10}}{1-F_{10}}}. \tag{2.5}$$

As $\alpha \to \propto$ (2.5) becomes (2.4) and $\vartheta_0 = 1$. We note that the pressure - diameter relation is complex.

Viscoelasticity is included by treating the latex as a voigt solid in the manner of Klip [2] and Gerrard [1]. Measurement of the logarithmic decrement of free oscillations showed the ratio of imaginary to real Young's moduli was 0.003 times the angular frequency over a range of frequencies. Viscoelasticity was included over the whole range of frequencies as explained by Gerrard [1].
The calculated wave speed for waves 1 and 2 are shown in Figure 2 as a function of α.

Womersley's theory was applied, Gerrard [1], to a freely suspended latex tube attached to a rigid tube in which a piston oscillated. The far end of the tube was closed and fixed. Far from the ends of a very long (34 m) tube this theory and the measured values were found to agree. End effects were modelled by the addition of an exponentially decaying entrance length. We will see later that in reality the situation in which waves 1 and 2 are both present may well be much more complicated.

Table 2.

Waves 1 → Waves II in a freely suspended tube.
Tube $x = 0$ to L divide into intervals of Δx
Let $L/\Delta x = Q$
Consider waves arriving at x, t and $^{x}/_{\Delta x} = P$

<u>1st Situation</u> Waves II arriving at x, t from the left

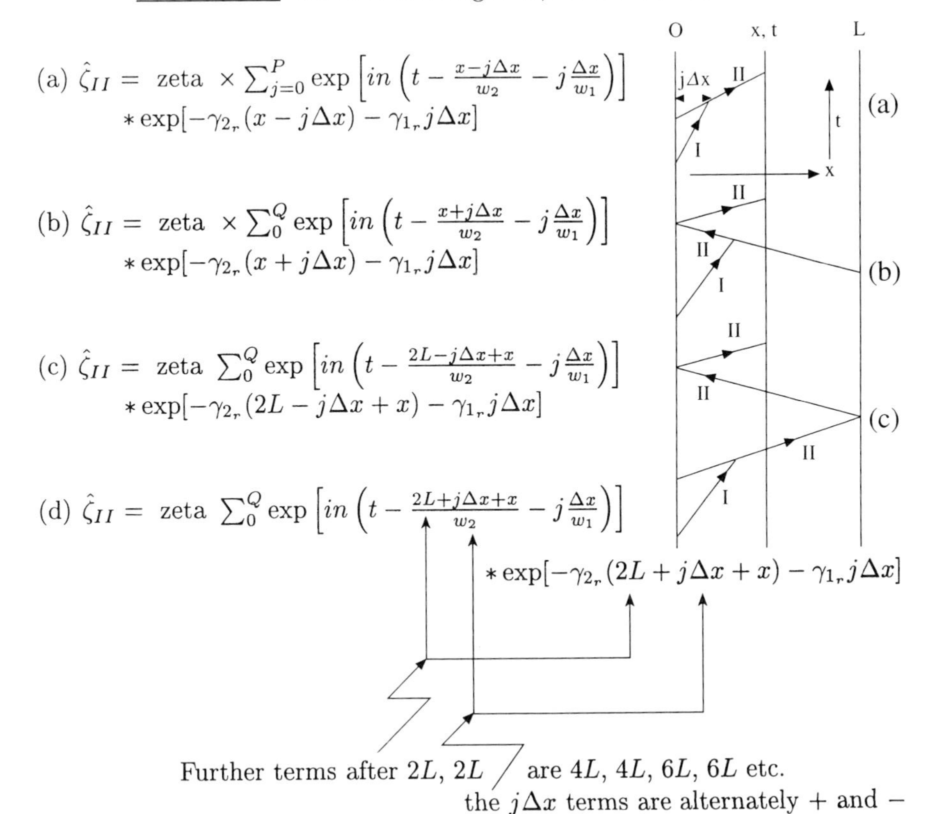

(a) $\hat{\zeta}_{II} = \text{zeta} \times \sum_{j=0}^{P} \exp\left[in\left(t - \frac{x - j\Delta x}{w_2} - j\frac{\Delta x}{w_1}\right)\right] * \exp[-\gamma_{2_r}(x - j\Delta x) - \gamma_{1_r} j\Delta x]$

(b) $\hat{\zeta}_{II} = \text{zeta} \times \sum_{0}^{Q} \exp\left[in\left(t - \frac{x + j\Delta x}{w_2} - j\frac{\Delta x}{w_1}\right)\right] * \exp[-\gamma_{2_r}(x + j\Delta x) - \gamma_{1_r} j\Delta x]$

(c) $\hat{\zeta}_{II} = \text{zeta} \sum_{0}^{Q} \exp\left[in\left(t - \frac{2L - j\Delta x + x}{w_2} - j\frac{\Delta x}{w_1}\right)\right] * \exp[-\gamma_{2_r}(2L - j\Delta x + x) - \gamma_{1_r} j\Delta x]$

(d) $\hat{\zeta}_{II} = \text{zeta} \sum_{0}^{Q} \exp\left[in\left(t - \frac{2L + j\Delta x + x}{w_2} - j\frac{\Delta x}{w_1}\right)\right] * \exp[-\gamma_{2_r}(2L + j\Delta x + x) - \gamma_{1_r} j\Delta x]$

Further terms after $2L$, $2L$ are $4L$, $4L$, $6L$, $6L$ etc.
the $j\Delta x$ terms are alternately + and −

Summation of a, b, c, d, etc. until further contributions are negligible.

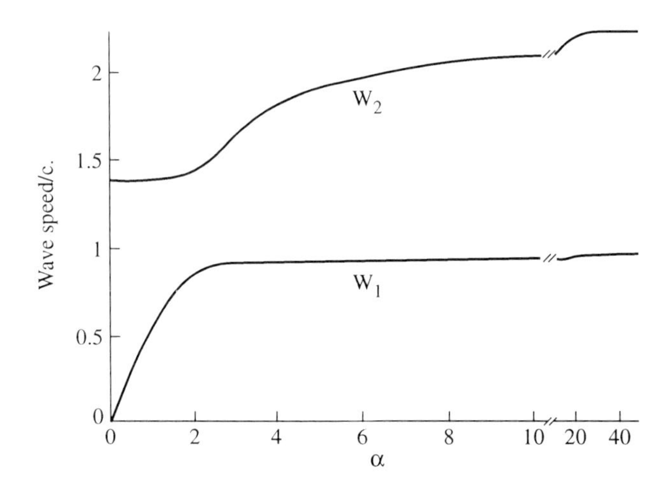

Figure 2. Wave speeds as a function of the Womersley parameter α

Table 3.

2nd Situation Waves II arriving at x, t from the right.

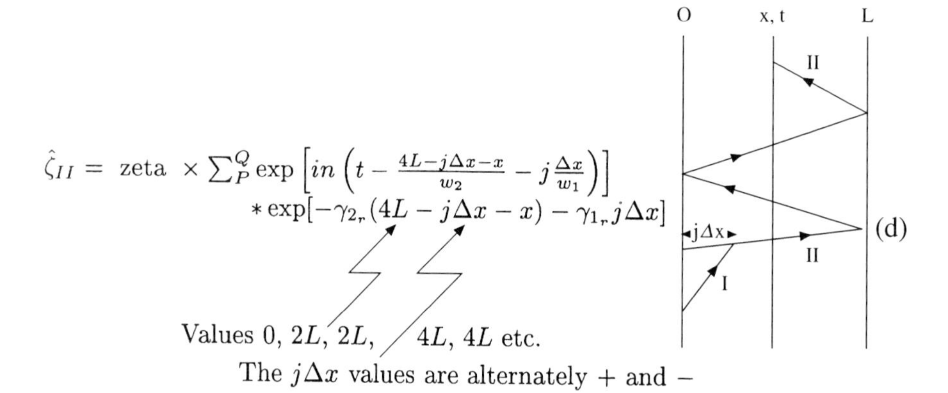

Summation is performed as in the first situation until further contributions are negligible.

In the initial motion as the piston moves forward fluid is moved into the latex tube which expands. Waves 1 progress down the tube and are continually produced as the piston oscillates. Initially, there are no waves 2: these are produced by the diminution of the length of an element due to its radial expansion. Waves 2 grow in amplitude from zero to reach a steady amplitude some distance down the tube. This production of waves 2 from waves 1 must continually occur at each element of the tube as wave 1 advances. The initial amplitude of wave 2 at each section we call zeta. Table 2 considers the situation in which waves 2 arrive at x, t travelling from left to right after being produced by waves 1 originating at $x = 0$. In Table 3 waves arriving at x, t travelling from the right are considered. Just as waves 1 produce wave 2, waves 1 are also produced by waves 2 and so on, so that the complete wave motion is the summation of a set of a doubly infinite series. We conclude that the full treatment of waves in tubes free to move longitudinally is extremely complicated. Freely moving tubes, however, have no cardiovascular application.

3 Waves in a tethered tube

In a tethered tube only waves 1 are present. The general arrangement of the experiment is shown in Figure 3. The circumferential strain of the latex tube was measured by mercury strain gauges connected to a high gain D.C. amplifier: these were applied to the tube as in Figure 4.

Oscillatory and single stroke piston motions were employed. These motions are shown in Figures 5 and 6. Figure 7 shows multiple traces from a strain gauge at the centre of the tube when pulses pass down the tube. The arrival of the first and second reflections can be seen. At later times the reflected pulses are lost in the background.
Figures 8 and 9 show that the pulse speed and oscillating wave speed are equal to 14 m/s. Because of the slower and reduced amplitude of the first cycle of oscillation the distance-time history of this cycle could be followed, see Figure 10. The analysis of the wave and its reflections shows that

$$\hat{\zeta}/R = C_1 \exp(-\gamma z) + C_2 \exp(\gamma z) \tag{3.1}$$

and

$$\hat{p} = kC_1 \exp(-\gamma z) + kC_2 \exp(\gamma z) \tag{3.2}$$

in which the linear value of k is used

$$\hat{p} = k\xi/R \text{ with } k = \frac{8}{3}\rho w_0^2 \tag{3.3}$$

w_0 is the inviscid wave speed $(Eh/2\rho R)^{1/2}$.

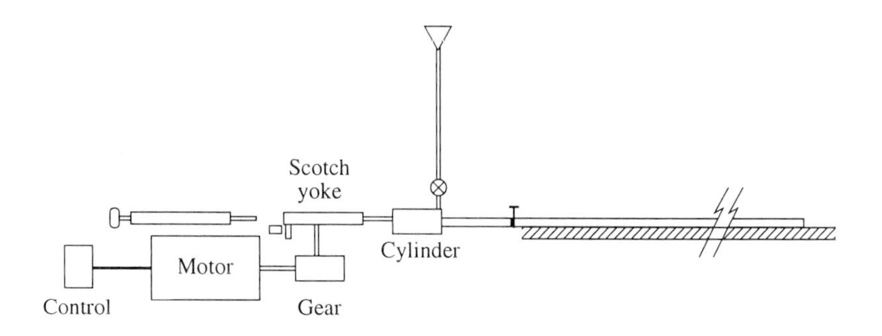

Figure 3. General arrangement of the experiment

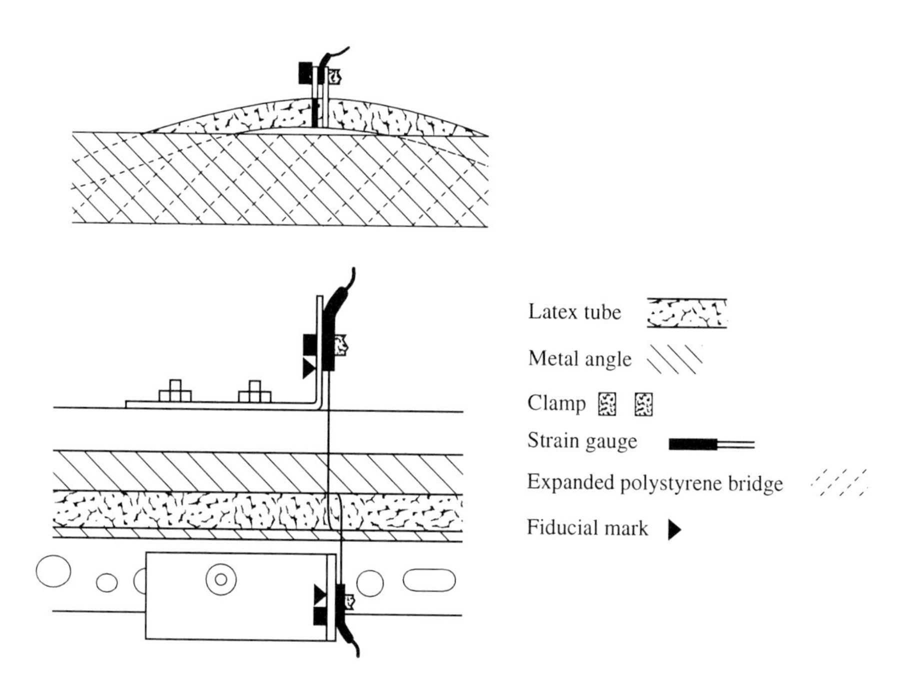

Figure 4. Mercury strain gauge application

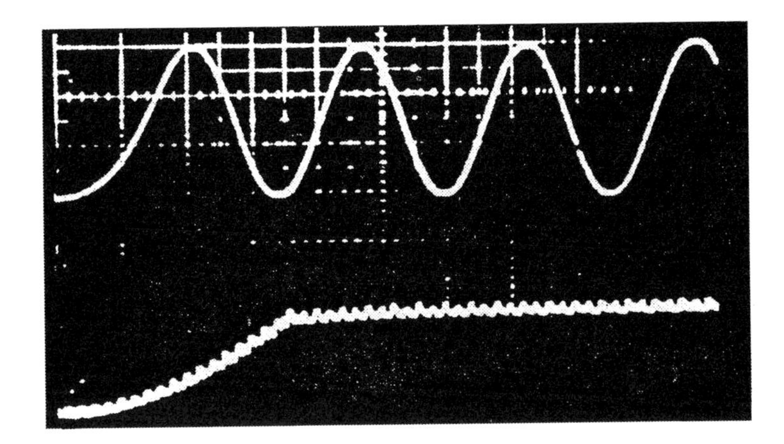

Figure 5. Start of oscillations at 4 Hz frequency. Top trace piston amplitude. Bottom trace motor speed

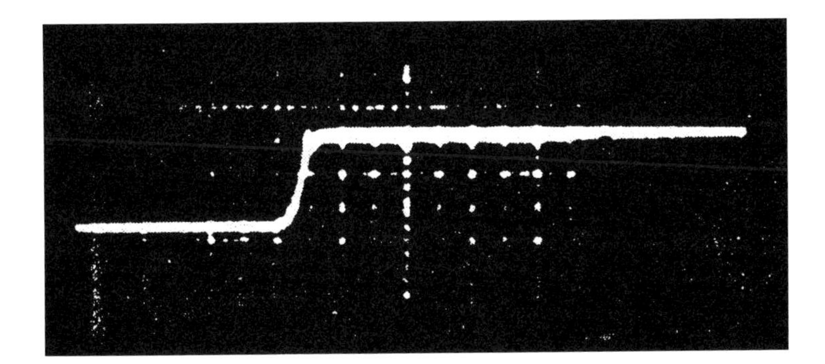

Figure 6. Piston position as a function of time when driven by the spring loaded trigger

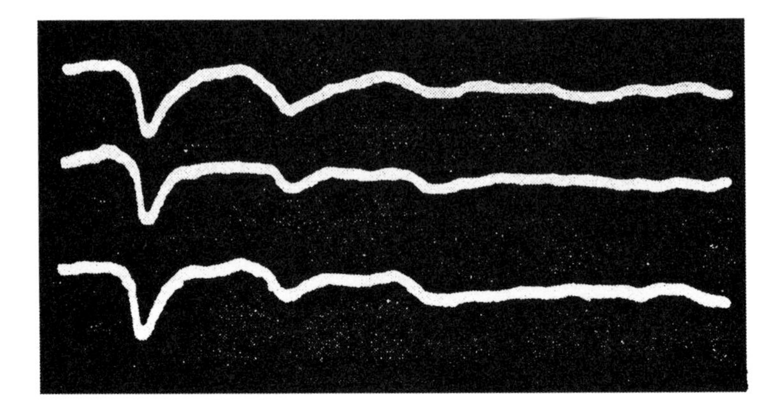

Figure 7. Multiple traces produced from a strain gauge at $z = 3$ m. Time base 200 ms/div

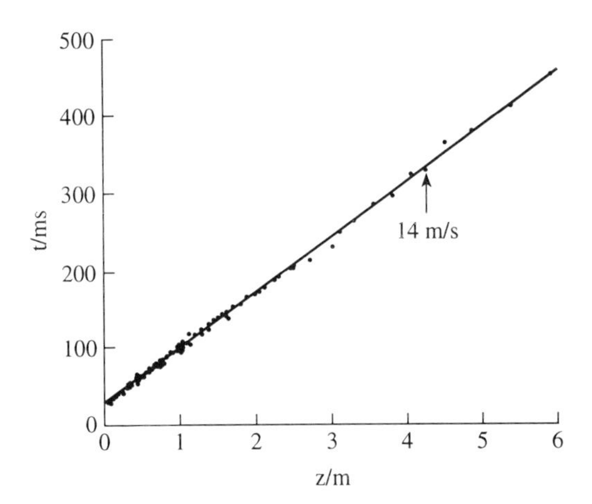

Figure 8. Arrival times of pulses

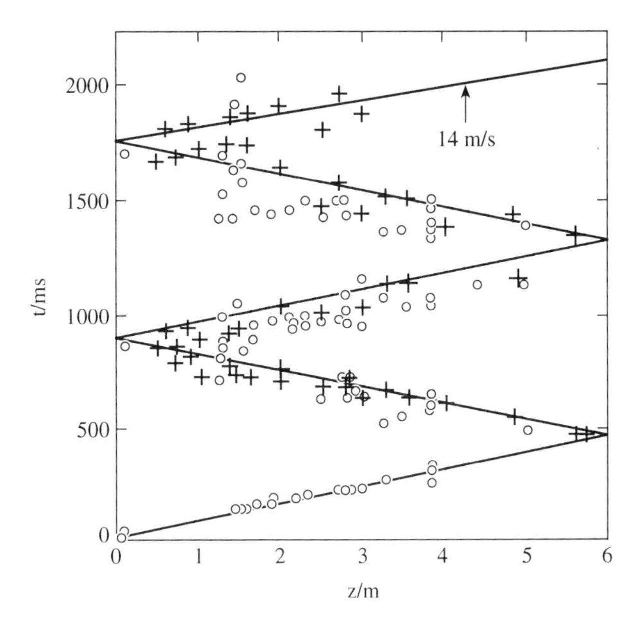

Figure 9. Arrival times of pulses (+) and of the start of oscillations (0)

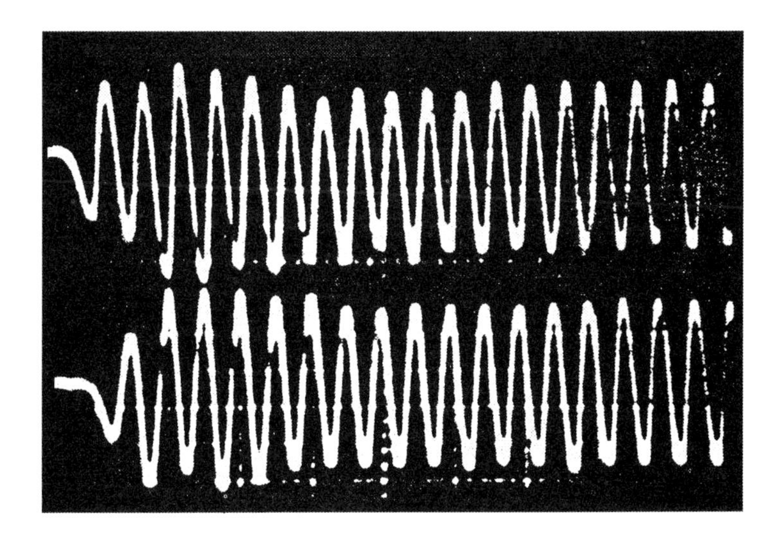

Figure 10. Simultaneous oscillogram from $z = 3.14$ and 4.50 m. Graticule divisions 2 V and 500 ms. Frequency 4 Hz

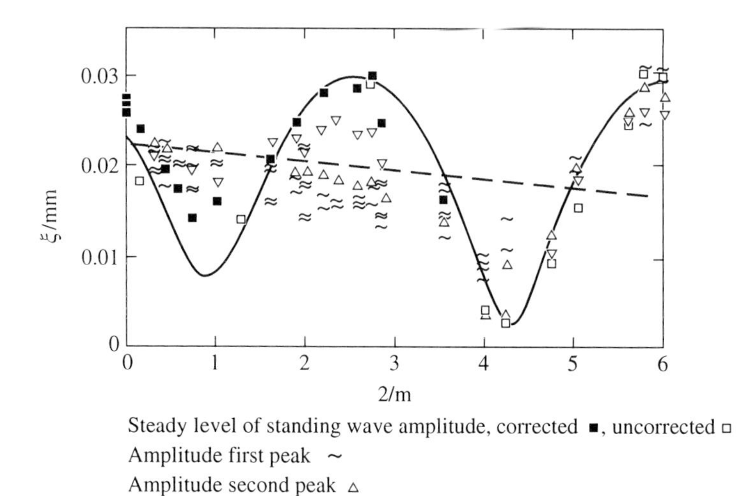

Figure 11. Amplitude of tube wall oscillations as a function of distance along the tube at frequency = 2 Hz

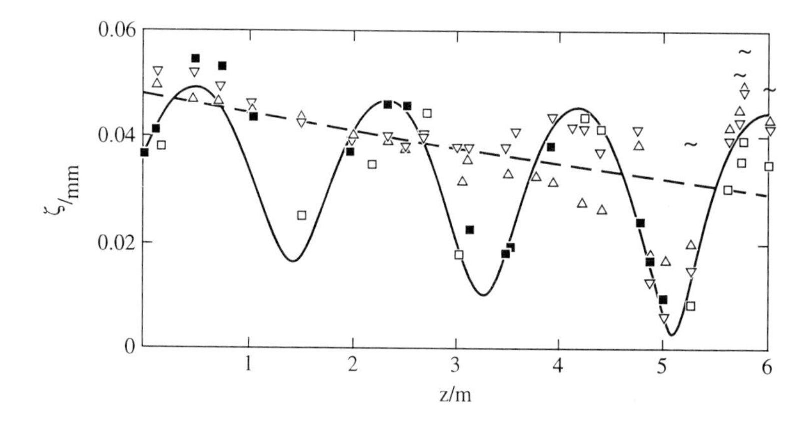

Figure 12. As Figure 11 except frequency 4 Hz

The boundary conditions are: at the piston end the volume flow rate matches that of the piston oscillation; at the closed end, $z = L$, the work of Lighthill [4] and [5] and Pedley [6] shows that

$$C_1 \exp(-\gamma L) = C_2 \exp(\gamma L). \tag{3.4}$$

The radial oscillation is therefore given by

$$\hat{\xi}/R = \frac{MV_p}{\gamma k} \frac{(\exp(\gamma(2L - z)) + \exp(\gamma z))}{1 - \exp(2\gamma L)} \tag{3.5}$$

with $M = -i\rho n V_p/(1 - F_{10})$ where V_p is the piston velocity and

$$F_{10} = Z J_1(\alpha i^{3/2})/(\alpha i^{3/2} J_o(\alpha i^{3/2}))$$

where J_0 and J_1 are Bessel functions of order 0 and 1.

In Figures 11 and 12 the results of the analysis are compared with the measured values. The first three peaks of the oscillation and the steady level were measured. When the first peak is not affected by reflection from $z = L$ the peaks at 2 and 4 Hz respectively attenuate according to

$$\hat{\xi} = 0.0223 \exp(-4.818 \times 10^{-4} z)\text{mm}$$

$$\text{and} \tag{3.6}$$

$$\hat{\xi} = 0.048 \exp(-8.804 \times 10^{-4} z)\text{mm}.$$

These curves are included in Figures 11 and 12. In the regions of smaller z where the first peak lay close to the curve a correction to the steady level was made by multiplication of the ξ value by the ratio of the measured ξ to the Equation (3.6). The second and third peaks are affected earlier by reflections. We conclude that the analysis agrees very well with the measured values of the standing wave amplitude.

Bibliography

1. Gerrard, J.H. (1985). An experimental test of the theory of waves in fluid-filled deformable tubes. *J. Fluid Mech.*, **156**, pp. 321–347.

2. Klip, W., van Loon, P., and Klip, D.A. (1967). Formulas for phase velocity and damping of longitudinal waves in thick walled viscoelastic tubes. *J. Appl. Phys.*, **38**, p. 3745.

3. Taylor, L.A., and Gerrard, J.H. (1977). Pressure-radius relationships for elastic tubes and their application to arteries. Part 1: Theoretical relationships, and Part 2: A comparison of theory and experiment for a rubber tube. *Med. and Biol. Eng. and Computing*, **15**, pp. 11–21.

4. Lighthill, M.J. (1973). *Mathematical Biofluiddynamics, Soc. Indust. and Appl. Math., Philadelphia*, **12**.

5. Lighthill, M.J. (1978). *Wavcs in Fluids, C.U.P.*, **2**.

6. Pedley, T.J. (1980). *The Fluid Mechanics of Large Blood Vessels, C.U.P.*, **2**.

Study of Flow Stability Using Direct Numerical Simulation of the Disturbance Equations

A. Spentzos and D. Drikakis[1]

Department of Mechanical Engineering, University of Manchester Institute of Science and Technology

Abstract

The present study deals with the flow stability in suddenly-expanded channels. The stability analysis is performed by numerically solving the linearised disturbance equations rather than the full set of the Navier-Stokes equations (see for example in [1]). The kinetic energy of the disturbances is employed as stability criterion for calculating the eigenvalue corresponding to the least stable mode and, subsequently, detecting the stability regimes. The computations reveal that for Reynolds numbers in the stability regime, an initial random flow perturbation is quickly attenuated. On the contrary, in the case of unstable flows the perturbations continue to exist as the time progresses, and concentrate in the regions of the flow field where instabilities appear.

1 Introduction

Incompressible flows in suddenly-expanded channels are encountered in various physiological flows for example, flows through prosthetic devices and aortic stenoses. At certain Reynolds numbers these flows present instabilities which may lead to bifurcation, unsteadiness, and chaos. Study of such flows can improve our understanding of hydrodynamic stability and the mechanism of laminar-to-turbulent flow transition.

Past experimental studies of low Reynolds number flows in symmetric suddenly-expanded channels have been performed by Chedron et al. [2] and Fearn et al. [3]. Those studies revealed a flow instability beyond a certain Reynolds number, which was appeared in the form of asymmetric separation between the upper and lower channel's walls. More recently, the same instability was also predicted by numerically solving the Navier-Stokes equations [1, 4, 5]. Moreover, in reference [1] the effects of the discretisation scheme and expansion ratio on the flow instability were also investigated.

Another approach is employed here for the stability analysis of such flows. This is based on the numerical solution of the linearised disturbance equations in

[1]Corresponding author, drikakis@umist.ac.uk.

conjunction with an energy stability criterion for calculating the eigenvalue corresponding to the least stable mode. The advantages of such an approach, compared to the direct numerical simulation of the Navier-Stokes equations, is that it is less computing intensive, especially if stablity analysis of three-dimensional flows is of interest, and can also be used to quantitatively examine the effects of disturbances on the flow stability. A similar procedure was also proposed in the past by Shapira et al. [6] and applied to detect the critical Reynolds number of suddenly-expanded channel flows. The ultimate objective of the present research is to develop a computational stability analysis procedure, based on the disturbance equations, for two- and three-dimensional flows. However, as a first step towards the above objective the prediction of the flow stability regimes as well as the evolution of disturbances in two-dimensional suddenly-expanded flows, are investigated.

2 Computational approach

The starting point of the present approach is the decomposition of the velocity components, U and V, and pressure, P, into the base-flow, (U_b, V_b, P_b), and disturbance components, (u, v, p):

$$(U, V, P) = (U_b, V_b, P_b) + (u, v, p). \tag{2.1}$$

Inserting the above decomposition into the Navier-Stokes equations and subtracting the base-flow field equations from the new equations, the linear stability equations are obtained:

$$\frac{\partial u}{\partial x} + \frac{\partial v}{\partial y} = 0 \tag{2.2}$$

$$\frac{\partial u}{\partial t} + U_b\frac{\partial u}{\partial x} + V_b\frac{\partial u}{\partial y} + u\frac{\partial U_b}{\partial x} + v\frac{\partial U_b}{\partial y} = -\frac{\partial p}{\partial x} + \frac{1}{Re}\left(\frac{\partial^2 u}{\partial x^2} + \frac{\partial^2 u}{\partial y^2}\right) \tag{2.3}$$

$$\frac{\partial v}{\partial t} + U_b\frac{\partial v}{\partial x} + V_b\frac{\partial v}{\partial y} + u\frac{\partial V_b}{\partial x} + v\frac{\partial V_b}{\partial y} = -\frac{\partial p}{\partial y} + \frac{1}{Re}\left(\frac{\partial^2 v}{\partial x^2} + \frac{\partial^2 v}{\partial y^2}\right). \tag{2.4}$$

The solution of the above equations in a time-dependent fashion provides information regarding the evolution of the flow disturbances. However, in order to detect the flow stability regimes a stability criterion needs to be defined. In the present study the eigenvalue corresponding to the least stable mode, b_{cr}, is used as a stability criterion. This is calculated by:

$$b_{cr} = \frac{1}{2\Delta t} ln \frac{E(t + \Delta t)}{E(t)} \tag{2.5}$$

where Δt is the time step and $E(t)$ is the total kinetic energy of the disturbances:

$$E(t) = \int_A (u^2 + v^2) dA. \tag{2.6}$$

According to above stability criterion, a stable or unstable flow will be defined according to the sign of the eigenvalue, b_{cr}. A positive value of b_{cr} will correspond to an unstable flow, whereas a negative value corresponds to a stable one.

Although the disturbance equations are simpler than the full set of the Navier-Stokes equations, the problem of coupling the continuity with the momentum equations still remains. In the present work such a coupling is obtained via the artificial compressibility formulation according to which a pseudo-time derivative is added to the continuity equation. For steady flows this derivative, as well as the time derivatives in the momentum equations, will eventually become zero when the steady state is reached. However, since we are interested in the time accurate solution of the disturbance equations, pseudo-time derivatives should also be added to the momentum equations in order to obtain a coupling at the pseudo-time level. Therefore, the equations are written as:

$$\frac{1}{\beta}\frac{\partial p}{\partial \tau} + \frac{\partial u}{\partial x} + \frac{\partial v}{\partial y} = 0 \tag{2.7}$$

$$\frac{\partial u}{\partial t} + \frac{\partial u}{\partial \tau} + U_b\frac{\partial u}{\partial x} + V_b\frac{\partial u}{\partial y} + u\frac{\partial U_b}{\partial x} + v\frac{\partial U_b}{\partial y} = -\frac{\partial p}{\partial x} + \frac{1}{Re}(\frac{\partial^2 u}{\partial x^2} + \frac{\partial^2 u}{\partial y^2}) \tag{2.8}$$

$$\frac{\partial v}{\partial t} + \frac{\partial v}{\partial \tau} + U_b\frac{\partial v}{\partial x} + V_b\frac{\partial v}{\partial y} + u\frac{\partial V_b}{\partial x} + v\frac{\partial V_b}{\partial y} = -\frac{\partial p}{\partial y} + \frac{1}{Re}(\frac{\partial^2 v}{\partial x^2} + \frac{\partial^2 v}{\partial y^2}) \tag{2.9}$$

where τ denotes the pseudo-time and β is the artificial compressibility parameter. Convergence at the pseudo-time level is achieved here by a first-order Euler scheme. Fourth-order dicretisation is used for the spatial derivatives. The results showed that the stability regimes are correctly predicted using the above approach, but various numerical issues which may affect the convergence of the solution and/or the stability analysis are currently investigated by the authors. The numerical issues under investigation include implementation of less dissipative schemes such as spectral methods and high-order temporal discretisation.

3 Results

In the present study, the flow in a suddenly-expanded channel with an expansion ratio 1 : 2 was considered. One of the reasons for selecting this flow case is the existence of a large amount of information regarding the stability of this flow, on the basis of the numerical solution of the Navier-Stokes equations [1, 4, 5]. Results for various expansion ratios and for several Reynolds numbers can be found in [1]. The flow instability appears in the form of an asymmetric separation between the upper and lower walls of the channel. The flow remains symmetric at very low Reynolds numbers with separation lengths of equal length on both channel walls. With increasing the Reynolds number the separation length increases too, and at a critical value asymmetries begin to appear: one recirculation region grows while the other shrinks, and this asymmetry increases with increasing the Reynolds number [1]. According to the results of reference [1], the critical Reynolds number for the symmetry-breaking bifurcation, based on the upstream channel height and maximum upstream velocity, was close to 216.

In the present work the objective is not to re-compute this critical value, but to examine the propagation of the disturbances for Reynolds numbers corresponding to stable and unstable flows. The investigation was performed for Reynolds numbers, Re, equal to $50, 100, 400$ and 750, respectively. For $Re = 50$ and 100 the flow is stable, whereas for $Re = 400$ and 750 the flow becomes unstable. In order to obtain the base-flow solution, the Navier-Stokes equations are initially solved in the half channel by applying symmetry conditions along the channel's centre-line. The base-flow solution for $Re = 100$ is shown in Figure 1. In the same figure the initial random field for the velocity disturbance

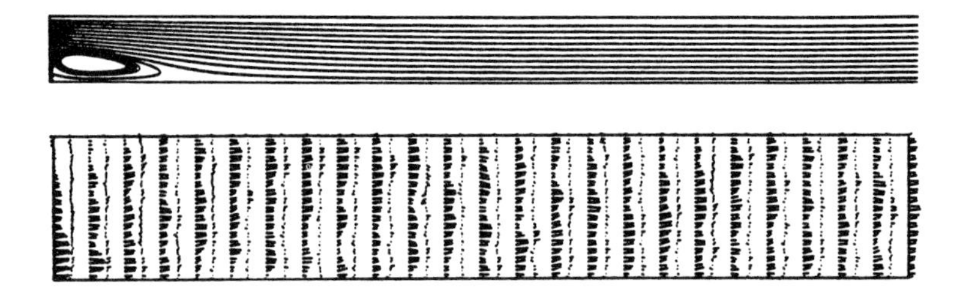

Figure 1. Base-flow for the half channel at $Re = 100$ (upper graph) and random initial field of disturbances (lower graph)

components is also shown. The same random field was given as initial condition for all cases considered here. At the inlet the disturbances were equal to zero and at the outlet the gradients of the disturbance quantities in the $x-$direction were also set equal to zero. No slip boundary conditions were applied for the velocity disturbances on the walls, while the gradient of the pressure disturbance normal to the wall was also set equal to zero. The variation of b_{cr} with time at $Re = 100, 400$ and 750 is shown in Figure 2. The b_{cr} clearly converges to a positive value for $Re = 400$ and 750, thus indicating an unstable flow, while it shows a periodic-like variation for $Re = 100$ retaining, however, always negative values, thus indicating a stable flow. The periodic-like variation of b_{cr} was also observed for other Reynolds numbers corresponding to stable flows, but the values of b_{cr} were always below zero for those cases, thus not affecting the final results. This convergence behaviour of b_{cr} could be attributed to various numerical issues, including grid size as well as spatial and temporal discretisation. These issues are currently under investigation. In the present study the grid-cell aspect ratio was equal to $1:5$ for all Reynolds numbers considered here.

The evolution of the velocity disturbances at $Re = 50, 100, 400$ and 750 is shown in Figures 3, 4, 5 and 6, respectively. Following Shapira et al. [6], the maximum length of the channel, L, was defined as function of the Reynolds number: $L = (8 + 0.08 \cdot Re)d$, where d is the upstream channel diameter. However, in order to make comparable the plots of Figures 3, 4, 5 and 6, the length was normalised by a certain factor and, therefore, the results in these figures should be considered from the qualitative point of view. For Reynolds numbers in the stable regime, the computations reveal that the disturbances propagate along the channel and attenuate after certain time. The lower the Reynolds number is, the faster attenuation of the disturbances is achieved. However, it is interesting to

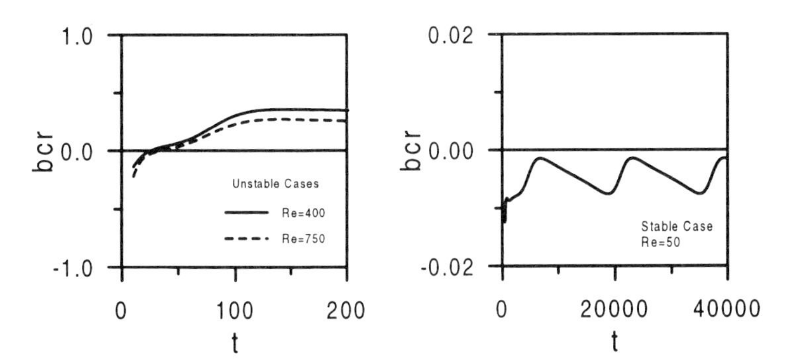

Figure 2. Evolution of the energy stability criterion for unstable flows at Re=400 and 750 (left graph), and for a stable flow at Re=50 (right graph)

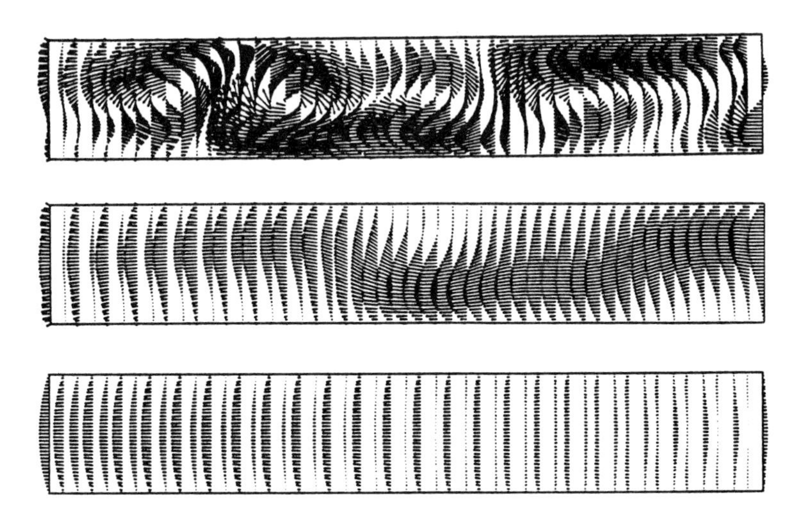

Figure 3. Vectors of the velocity disturbances at different time instants for $Re = 50$

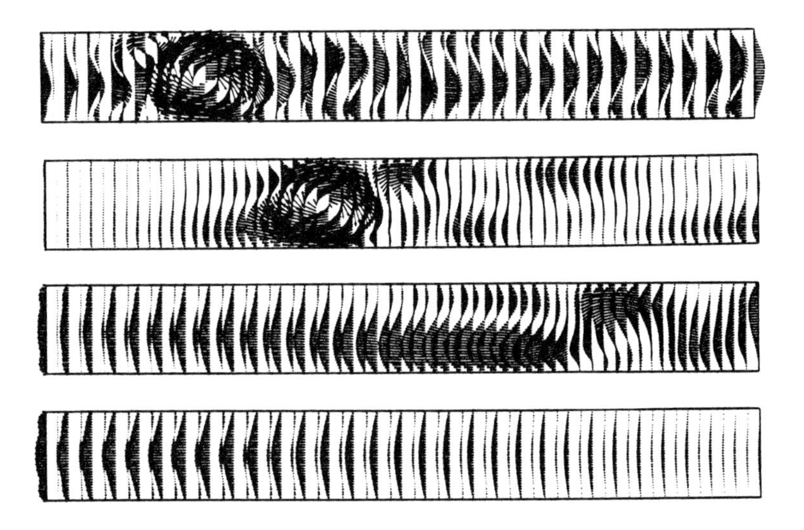

Figure 4. Vectors of the velocity disturbances at different time instants for $Re = 100$

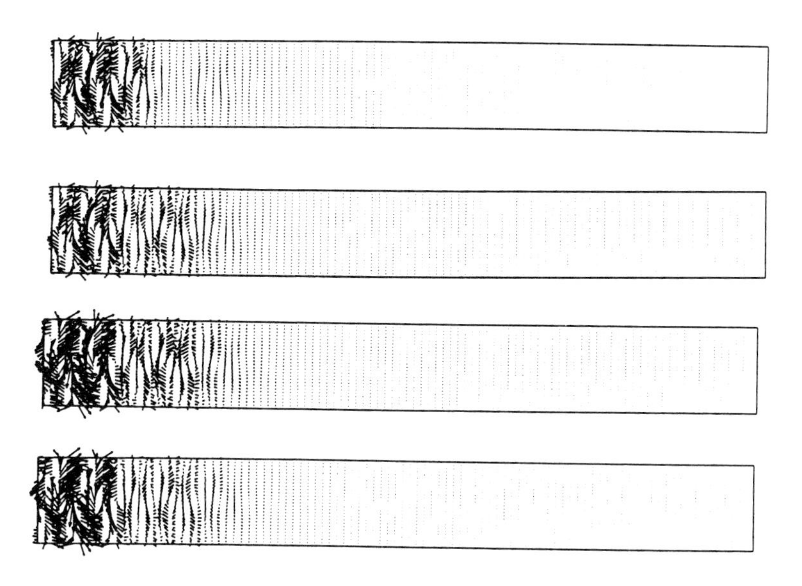

Figure 5. Vectors of the velocity disturbances at different time instants for $Re = 400$

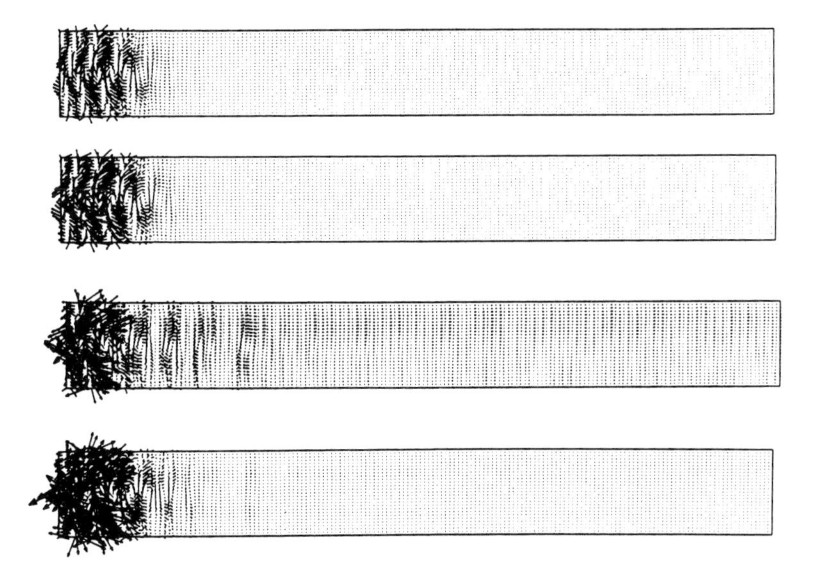

Figure 6. Vectors of the velocity disturbances at different time instants for $Re = 750$

observe that this is not the case at higher Reynolds numbers corresponding to unstable flows (Figures 5 and 6). In these cases the initial field of the disturbances does not die out, but concentrates in the regions where instabilities appear. It was actually found that the size of the region in which the highest values of the disturbances kinetic energy exist, fits very closely to the length of the channel where asymmetric separation appears (see [1]). Moreover, it is seen (Figure 5) that for $Re = 400$ the evolution of the flow disturbances is very much similar with that of $Re = 750$. Therefore, it seems that for the unstable flows considered here there is a high energy concentration of the disturbances in flow regimes where instabilities appear.

4 Conclusions

Stability analysis of the flow in a suddenly-expanded channel with an expansion ratio $1:2$ was performed via the numerical solution of the disturbance equations in conjunction with an energy stability criterion. Such a computational approach was first proposed by Shapira et al. [6], while the present study focused on the propagation of the disturbances revealing some new features regarding the behaviour of the disturbed field in stable and unstable flows.

The energy stability criterion correctly predicted the stable/unstable flow regimes for the Reynolds numbers considered in this study. The computations revealed that for stable flows the disturbances are propagating inside the channel and finally die out after certain time. The attenuation of the disturbances is faster when reducing the Reynolds number, i.e. when we are moving away from the critical Reynolds number that the instability begins to appear. In contrast to the above, in unstable flows the disturbances do not attenuate, but they are concentrated in flow regimes where instabilities appear.

The effects of the numerical accuracy on the stability predictions are currently investigated by implementing high-order spectral methods for the solution of the disturbance equations. Moreover, research regarding the investigation of the flow stability at higher Reynolds numbers is under way.

Bibliography

1. Drikakis, D. (1997). Bifurcation phenomena in incompressible sudden-expansion flows. *Physics of Fluids, 9(1)*, 76–87.

2. Chedron, W., Durst, F. and Whitelaw, J.H. (1978). Asymmetric flows and instabilities in symmetric ducts with sudden expansion. *J. Fluid Mech., 84*, 13–31.

3. Fearn, R.M., Mullin, T. and Cliffe, K.A. (1990). Nonlinear flow phenomena in a symmetric sudden expansion. *J. Fluid Mech., 211*, 595–608.

4. Alleborn, N., Nandakumar, K., Raszillier, H. and Durst, F. (1997). Further contributions on the two-dimensional flow in a sudden expansion. *J. Fluid Mech.*, *330*, 169–188.

5. Battaglia, F., Tavener, S.J., Kulkarni, A.K. and Merkle, C.L. (1997). Bifurcation of low Reynolds number flows in symmetric channels. *AIAA J.*, *35(1)*, 99–105.

6. Shapira, M., Degani, D. and Weihs, D. (1990). Stability and existence of multiple solutions for viscous flow in suddenly enlarged channel. *Computer Fluids*, *18*, 239–258.

Turbulent Flow in an Arterial Stenosis

S.G. Sajjadi and Y. Feng

Centre for Computational Fluid Dynamics and Turbulence, University of Salford

Abstract

A theoretical study for pulsatile flow in an elliptical tube of slowly varying cross-section is carried out. The main objective is to determine the flow conditions which may be produced by the presence of stenosis in arteries. For simplicity it is assumed that blood is Newtonian and the geometry of the stenosis is introduced by specifying the change in area of cross-section of the stenosed artery with axial distance. Exact solutions for axial and secondary velocities together with pressure gradient are derived in terms of Mathieu functions for a flow in which volume flux is prescribed. An expression for shear stress on the wall of the tube is presented. Results for axial velocity, impedance and shear stress are obtained for constricted tube. It is concluded that the characteristics of the flow are affected by the percentage change of stenosis, the geometry of stenosis as well as the frequency of oscillation. Results from numerical models are also included which shows a good level of agreement with experimental data.

1 Introduction

Cardiovascular diseases for example acute myocardial infection, congestive heart failure and stroke are still most frequent causes of death within the population of European Union (EU). To continue and accelerate the pace of the discoveries of the last decades and translate these findings into treatment, multidisciplinary cardiovascular research must be developed. Studies are best and most economically accomplished in a research setting combining the expertise of physicians and scientists with different backgrounds in basic and clinical cardiovascular research and in molecular genetics.

Although the initiating factor for the development of atherosclerosis has yet to be fully established it is believed that the development of localized arterial stenosis alters the flow and the nutrient supply to the wall at least in the neighbourhood of the constriction. Therefore, it is essential to understand the flow mechanism in the presence of a constriction, particularly the flow separation downstream of the constriction and the behaviour of turbulence in the recirculating region. This will then lead to a deeper understanding of the pathology of atherosclerosis.

It is further believed that haemodynamic factors may also play an important role in the initiation and progression of atherosclerosis. As yet no correlation between the haemodynamics and the initiating mechanism of the disease is fully established. However, the preferential distribution of the fully streaks at certain locations of low wall shear stress has indicated that shearing stress may possibly be responsible in the initiation of the disease. Furthermore, research indicates that high shear stress on the arterial walls can cause damage to the endothelium (Fry 1968). Although this could not explain the genesis of atherosclerosis, the relatively higher shear stress on the stenotic side could be an important contributing factor to the progressive growth of the stenotic plaque. Shearing stresses on the vessel wall may cause morphological changes to the endothelium. Although the specific role of wall shear stress in the atherosclerotic process has not been clearly established, our modelling study indicates that there is a distinct and significant difference in the elevated shear stresses on the opposite sides of the walls in an asymmetric stenosis (Sajjadi and Feng 1998). Our studies confirm that the wall with stenotic constriction experiences a higher wall shear stress than the other unprotruded side. The ratio of the peak shearing stress on the stenotic side to that on the unprotruded side of the arterial wall can be as high as 2.5 times, depending on the constriction percentage as well as Reynolds number.

As stresses are difficult, and in some cases impossible, to measure experimentally, it is thus, useful to be able to examine them in a mathematical model to gain further insight into this important aspect of haemodynamics. We can, for instance, study the effects of flow in curved arteries or at bifurcations on the distribution of wall shear stress, which in fact, will be the subject of our further study in this area.

Caro and Saffman (1965) have pointed out the importance of modelling arterial blood flows in pipes of elliptical cross-sections. Thus, the steady flow counterpart of the present problem for an incompressible viscous fluid in a slow varying elliptical tube in Cartesian coordinates has ben studied by Wild et al. (1977). More recently Mehrotra et al. (1985) have attempted to extend the analysis of Wild et al. (1977) to pulsatile flow. However, their solution for primary unidirectional velocity is incorrect due to omission of the important secondary velocities that arise due to variation in the cross-section. The present contribution, based on the work of Padmavathi and Rao (1992), investigates the same problem as that of Mehrotra et al. (1985) in elliptical coordinate system (ξ, η, z) using a perturbation analysis. An appropriate geometric perturbation parameter for the present problem is $\varepsilon = \ell/L$. Here $\ell = (a_0 + b_0)/2$, where a_0 and b_0 are semimajor and semiminor axes in the region of constant cross-section and L is a characteristic length along the tube in z-direction. Closed form solutions and their asymptotic representations are derived for the zeroth order axial velocity and ε order velocity components in ξ and η directions. The pressure drop, wall shear stress and impedance are calculated across the stenosis.

2 Formulation of the problem

Consider the unsteady flow of an incompressible viscous fluid of constant density ρ and kinematic viscosity ν in an elliptical tube of slowly varying cross-section. Let (ξ, η, z) be the elliptical cylindrical coordinates given by

$$x + iy = c\cosh(\xi + i\eta), \qquad z = z \tag{2.1}$$

with $c > 0, 0 \leq \eta \leq 2\pi, 0 \leq \xi \leq \infty$ and $-\infty < z < \infty$. At any given axial station, z =constant, the cross-section is given by $\xi = \xi_0(z)$, the semimajor axis is $a(z) = c\cosh\xi_0(z)$, the semiminor axis is $b(z) = c\sinh\xi_0(z)$, where $c^2 = a^2 - b^2$.

Upon introducing the non-dimensional variables

$$\begin{array}{lll} z = Lz^*, & (u, v, w) = (U_0\varepsilon u^*, U_0\varepsilon v^*, U_0 w^*) & \\ t = \omega t^*, & R = U_0\ell/\varepsilon, & p = \rho U_0^2 p^*/R\varepsilon \\ k = c/\ell, & \alpha^2 = \omega\ell^2/\nu & \end{array}$$

the Navier-Stokes equations governing the flow of a Newtonian fluid in elliptical coordinates become (after dropping the asterisks):

$$\frac{\partial}{\partial\xi}(fu) + \frac{\partial}{\partial\eta}(fv) + k^2 f^2 \frac{\partial w}{\partial z} = 0 \tag{2.2}$$

$$\alpha^2 k^2 \frac{\partial u}{\partial t} + R\varepsilon k\left[D_1 u + \frac{v}{f^2}D_2 f\right] = \frac{-k}{\varepsilon^2 f}\frac{\partial p}{\partial\xi} + \frac{1}{f}\left[\nabla^2(fu) - \frac{1}{f^4}\{D_3(fu) + D_4(fv)\}\right] \tag{2.3}$$

$$\alpha^2 k^2 \frac{\partial v}{\partial t} + R\varepsilon k\left[D_1 v - \frac{u}{f^2}D_2 f\right] = \frac{-k}{\varepsilon^2 f}\frac{\partial p}{\partial\eta} + \frac{1}{f}\left[\nabla^2(fv) - \frac{1}{f^4}\{D_3(fv) - D_4(fu)\}\right] \tag{2.4}$$

$$\alpha^2 k^2 \frac{\partial w}{\partial t} + R\varepsilon k D_1 w = -k^2\frac{\partial p}{\partial z} + \nabla^2 w. \tag{2.5}$$

Here (u, v, w) are velocity components in (ξ, η, z) directions, α is Womerslay parameter[1], R is the Reynolds number, the differential operators $D_1 ... D_4$ are given by

$$D_1 \equiv \frac{u}{f}\frac{\partial}{\partial\xi} + \frac{v}{f}\frac{\partial}{\partial\eta} + wk\frac{\partial}{\partial z}, \qquad D_2 \equiv u\frac{\partial}{\partial\eta} - v\frac{\partial}{\partial\xi},$$

$$D_3 \equiv \sinh(2\xi)\frac{\partial}{\partial\xi} + \sin(2\eta)\frac{\partial}{\partial\eta}, \qquad D_4 \equiv \sinh(2\xi)\frac{\partial}{\partial\eta} - \sin(2\eta)\frac{\partial}{\partial\xi}$$

and the Laplacian in this coordinate system is

$$\nabla^2 \equiv \frac{1}{f^2}\left[\frac{\partial^2}{\partial\xi^2} + \frac{\partial^2}{\partial\eta^2}\right] + k^2\varepsilon^2\frac{\partial^2}{\partial z^2}$$

[1]Commonly known as the frequency parameter.

where $f = \sqrt{[\cosh(2\xi) - \cos(2\eta)]/2}$.

Equations (2.3)–(2.5) are to be solved subject to the boundary conditions

$$u = v = w = 0 \qquad \text{on} \qquad \xi = \xi_0(z). \tag{2.6}$$

The non-dimensional flux across any given cross-section is defined by

$$Q = \int_0^{2\pi} \int_0^{\xi_0(z)} w f^2 \, d\xi \, d\eta = \pi \left(1 + e^{it}\right). \tag{2.7}$$

3 Perturbation and asymptotic analysis

Expanding the velocities and pressure in a perturbation series

$$(u, v, w, p) = \sum_{n=0}^{\infty} \varepsilon^n (u_n, v_n, w_n, p_n)$$

thence, upon substitution in Equations (2.2)–(2.5) and collecting like powers of ε, we obtain the following system:

$$\frac{\partial p_0}{\partial \xi} = \frac{\partial p_0}{\partial \eta} = 0, \qquad \alpha^2 k^2 \frac{\partial w_0}{\partial t} = \frac{1}{f^2}\left\{\frac{\partial^2}{\partial \xi^2} + \frac{\partial^2}{\partial \eta^2}\right\} w_0 - k^2 \frac{\partial p_0}{\partial z} \tag{3.1}$$

$$\frac{\partial}{\partial \xi}(f u_0) + \frac{\partial}{\partial \eta}(f v_0) + k^2 f^2 \frac{\partial w_0}{\partial z} = 0 \tag{3.2}$$

$$\alpha^2 k^2 \frac{\partial u_0}{\partial t} = \frac{1}{f}\left[\frac{1}{f^2}\left\{\frac{\partial^2}{\partial \xi^2} + \frac{\partial^2}{\partial \eta^2}\right\}(f u_0) - \frac{1}{f^4}\{D_3(f u_0) + D_4(f v_0)\}\right] - \frac{k}{f}\frac{\partial p_2}{\partial \xi} \tag{3.3}$$

$$\alpha^2 k^2 \frac{\partial v_0}{\partial t} = \frac{1}{f}\left[\frac{1}{f^2}\left\{\frac{\partial^2}{\partial \xi^2} + \frac{\partial^2}{\partial \eta^2}\right\}(f v_0) - \frac{1}{f^4}\{D_3(f v_0) - D_4(f u_0)\}\right] - \frac{k}{f}\frac{\partial p_2}{\partial \eta}. \tag{3.4}$$

If in Equation (3.1) we let

$$w_0 = w_{0S} + w_{0T} e^{it}, \qquad \frac{dp_0}{dz} = p_{0S} + p_{0T} e^{it}$$

the solutions that satisfies the boundary conditions (2.6) and the flux (2.7) are

$$w_{0S} = \frac{k^2}{8\cosh(2\xi_0)} p_{0S} [\cosh(2\xi) - \cosh(2\xi_0)][\cosh(2\xi_0) - \cos(2\eta)] \tag{3.5}$$

and

$$w_{0T} = \frac{k p_{0T}}{\beta^2}\left\{1 - 2\pi \sum_{m=0}^{\infty} \frac{A_0^{(2m)}}{L_{2m} Ce_{2m}(\xi_0, q)} Ce_{2m}(\xi, q) ce_{2m}(\eta, q)\right\}. \tag{3.6}$$

Here $Ce_{2m}(\xi, q)$ and $ce_{2m}(\eta, q)$ are the modified Mathieu and Mathieu functions which are related by the relationship

$$Ce_{2m}(\xi, q) = ce_{2m}(i\xi, q).$$

The expansion of $ce_r(\xi, q)$ for small $|q|$ is given by (Abramowitz and Stegun 1972)

$$ce_{2m}(\xi, q) = \cos(2m\xi) - q\left[\frac{\cos(m+1)\xi}{2(2m+1)} - \frac{\cos(m-1)\xi}{2(2m-1)}\right] + O(q^2) \tag{3.7}$$

whilst for $|\xi| \gg 1$ the asymptotic form for $Ce_{2m}(\xi, q)$ may be expressed as (Erdélyi et al. 1954)

$$Ce_{2m}(\xi, q) \sim P_{2m}\left(\tfrac{1}{2}\pi\right)^{-1/2} q^{-1/4} e^{-\xi/2} \cos\left(q^{1/2}e^{\xi} - \tfrac{1}{4}\pi\right) \tag{3.8}$$

where $q = \beta^2/4 = -i\alpha^2 k^2/4$ and

$$P_{2m} = \frac{ce_{2m}(0, q)ce_{2m}(\pi/2, q)}{A_0^{(2m)}}, \qquad A_0^{(2m)} = \frac{(m-1)!}{m!(2m-1)!}\left(\frac{q}{4}\right)^m.$$

Note that in (3.6) L_{2m} is given by

$$L_{2m} = \int_0^{2\pi} ce_{2m}^2(s, q)\, ds = \pi + O(q^2) \tag{3.9}$$

where the latter equality follows from integrating (3.7).

Substituting (3.7)–(3.9) into (3.6), we obtain the following closed form solution

$$w_{0T} = \frac{kp_{0T}}{\beta^2}\left\{1 - 2\sum_{m=0}^{\infty} \frac{(m-1)!}{m!(2m-1)!}\left(\frac{q}{4}\right)^m e^{-\frac{1}{2}(\xi-\xi_0)} \frac{\cos\left(q^{1/2}e^{\xi} - \frac{1}{4}\pi\right)}{\cos\left(q^{1/2}e^{\xi_0} - \frac{1}{4}\pi\right)}\right.$$
$$\left. \times \left[\cos(2m\eta) - q\left(\frac{\cos(m+1)\eta}{2(2m+1)} - \frac{\cos(m-1)\eta}{2(2m-1)}\right)\right]\right\}. \tag{3.10}$$

Having obtained a solution for w_0, we are in a position to calculate the pressure p_0 from the flux condition (2.7). Substituting (3.5) and (3.10) into (2.7), performing the double integration and equating the real and imaginary parts, we obtain

$$p_{0S} = \frac{-32\cosh(2\xi_0)}{k^2\sinh^3(2\xi_0)} \tag{3.11}$$

and

$$p_{0T} = \left(\frac{\beta}{k}\right)^2 \left\{\frac{\sinh(2\xi_0)}{2} - \sum_{m=0}^{\infty} \frac{\varpi^{(2m)} - \vartheta^{(2m)}}{m!(2m-1)!\cos\left(q^{1/2}e^{\xi_0} - \frac{1}{4}\pi\right)}\right\}^{-1} \tag{3.12}$$

where

$$\varpi^{(2m)} = (m-1)!\left(A_2^{(2m)}\xi_0 + \tfrac{1}{2}A_0^{(2m)}\right)A_0^{(2m)}, \qquad \vartheta^{(2m)} = A_0^{(2m)}A_2^{(2m)}\xi_0$$

and

$$A_{2n+2r}^{(2n)} = (-1)^r \frac{(2n)!}{r!(2n+r)!}\left(\frac{q}{4}\right)^r.$$

Similarly for the secondary velocities u_0 and v_0 we may write

$$u_0 = u_{0S} + u_{0T}e^{it}, \qquad v_0 = v_{0S} + v_{0T}e^{it}.$$

Upon substituting these in the equation of continuity (3.2) and equations of motion (3.3) and (3.4), we may obtain a similar type of expressions as those of (3.5) and (3.10). However, since our analysis is confined to mild constrictions, the magnitude of these secondary velocities are not very significant and do not affect the main conclusions to be drawn here[2]. Indeed, this has been amply confirmed by numerical simulations of flow in constricted tubes.

4 Results

In the real life situation, the stenosis that is developed in an artery has a random patchy raised spot which leads to an arbitrary shape. However, such a random shape cannót be incorporated in our present analysis and therefore we are forced to assume a particular shape for its geometrical representation. Hence, as we mentioned in Section 2, we have chosen

$$a(z) = c\cosh\xi_0(z), \qquad b(z) = c\sinh\xi_0(z) \tag{4.1}$$

for the stenosis considered in the present problem. This choice has been made in order not to violate the assumption of mild stenosis in our analysis. Hence, Equations (4.1) leads to 30% constriction where both major and minor axes are reduced. Thus, the mathematical representation of stenosis is identical to that of prescribing an equation on the boundary of an elliptic tube. Here we shall consider two different percentage of constrictions represented by (i) $\xi_0(z) = 2 - \frac{1}{2}e^{-z^2}$ and (ii) $\xi_0(z) = 1 - \frac{1}{2}e^{-z^2}$.

In human beings, the frequency parameter, α, varies from 0 to 10 in the resting condition, but may take higher values under conditions of exercise. However in the cases where $\alpha > 5$ the present asymptotic analysis is not valid, and hence our results will be confined to the cases where $0 \le \alpha \le 5$.

In Figure 1 we plot the axial velocity w_0 along the centerline of the tube for case (i) and (ii) above at time $t = 0$ for two values of frequency parameter $\alpha = 0$ and 5. As can be seen from this figure, the axial velocity is maximum at the point of peak stenosis ($z = 0$) for a given value of α and increases with

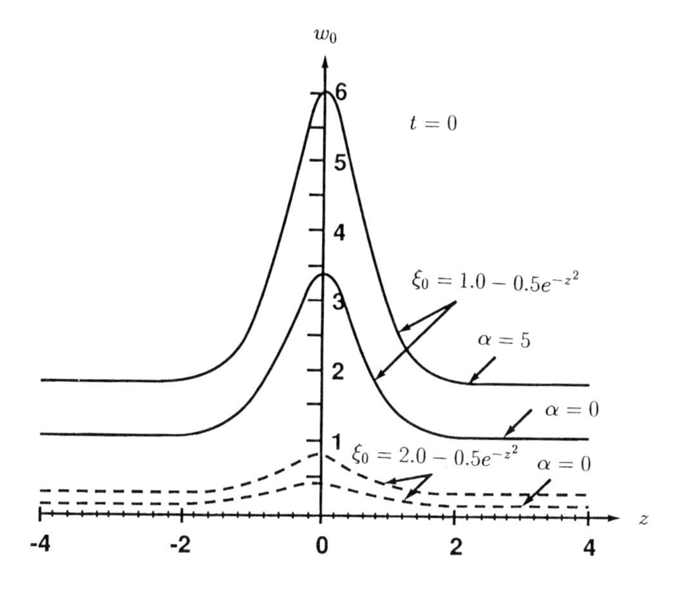

Figure 1. Axial velocity along the centerline

an increase in the percentage constriction in order to compensate the constant volume flux which is a direct consequence of the continuity Equation (2.2).

A useful quantity in the present calculation is the impedance I which is given by

$$I = \frac{\Delta p}{Q}. \tag{4.2}$$

Here Δp is the pressure drop across the stenosis and Q is the volume flux, i.e. the volume of blood flow. This quantity can be interpreted as the resistance offered to the flow by the stenosis. Figure 2 shows the variation of $|I/I_s|$, where I and I_s are impedances for pulsatile and steady flow cases respectively, with the frequency parameter α for different percentage constrictions (i) and (ii) at $z = 0$ and $t = 0$. It can be seen from this figure that $|I/I_s| \approx 1$ for small values of α and decreases with an increase in the percentage constriction as a direct consequence of the increased axial velocity w_0.

---

[2] u_0 and v_0 are only needed for determining the wall shear stress. The explicit expressions for these are not given here for brevity.

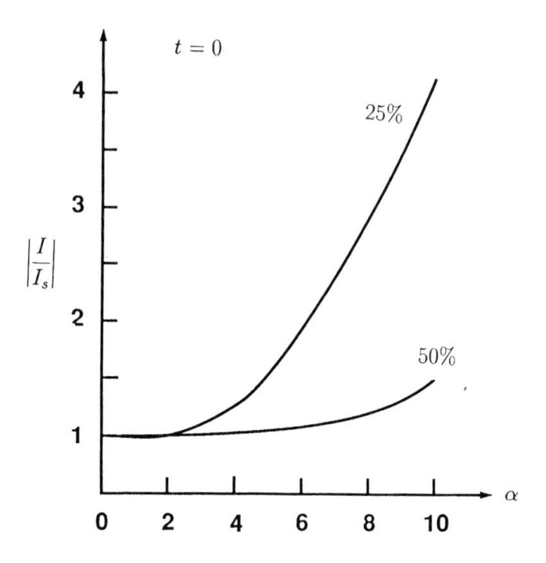

Figure 2. Magnitude of impedance at $z = 0$

The estimation of the wall shear stress is particularly important because of its possible role in the mass transfer rate across artery walls and atherogenesis. The wall shear stress may be expressed as

$$\tau_w = \left\{\tau_{\xi z}\left[1 - (d\xi_0/dz)^2\right] + (\sigma_{\xi\xi} - \sigma_{zz})(d\xi_0/dz)\right\}\left\{1 + (d\xi_0/dz)^2\right\}^{-1} \quad (4.3)$$

where $\tau_{\xi z}$ is the component of the shear stress in ξz-plane and $\sigma_{\xi\xi}$ and σ_{zz} are components of the normal stresses, in elliptical cylindrical coordinates. Figure 3 shows the wall shear stress across the stenosis at $t = 0$ for different values of frequency parameter α for geometry (i). As can be seen from this figure, the wall shear stress is maximum at the point of peak stenosis and has more than one point of extremum for a given value of α. Furthermore, the shear stress increases as the frequency parameter increases.

We shall now display some results obtained from numerical simulations. All these simulations are performed using a differential second moment closure model of turbulence developed by Sajjadi and co-workers, see for example Sajjadi and Aldridge (1995) or Sajjadi and Waywell (1997).

Figures 4(a)–4(c) display the streamlines for turbulent flow through arterial stenosis for Reynolds numbers 5000, 10000 and 15000, respectively. The results show high levels of turbulence is generated in the re-circulation region particularly for Reynolds number > 10000, Figures 4(b) and 4(c). Figure 4(d) shows the comparison between measurements ($\diamond$) (Deshpande and Giddens 1980) and computation (—) for the velocity along the axis dips under the fully developed

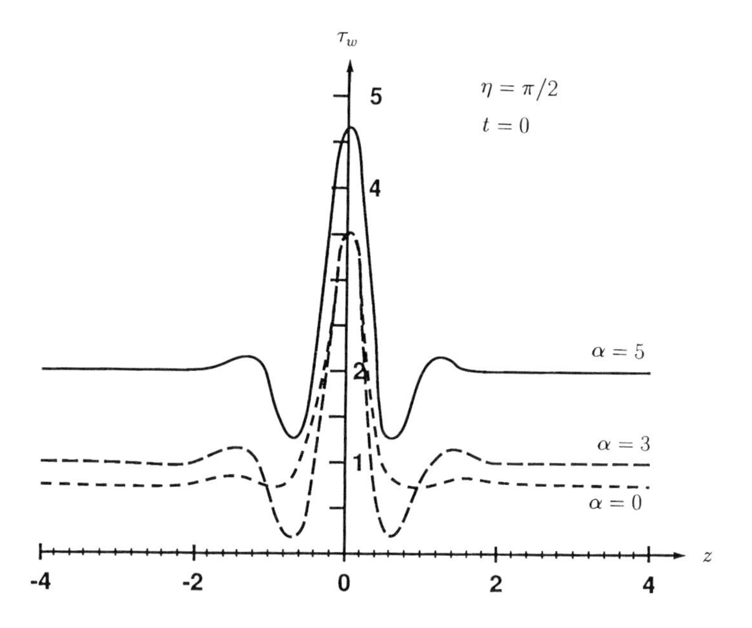

Figure 3. Wall shear stress at $t = 0$

value at Reynolds number 15000. This result bears an interesting similarity to the case of turbulent flow development in the entrance region of a pipe where the centre-line mean velocity reaches a maximum and then recedes to its final value as flow develops.

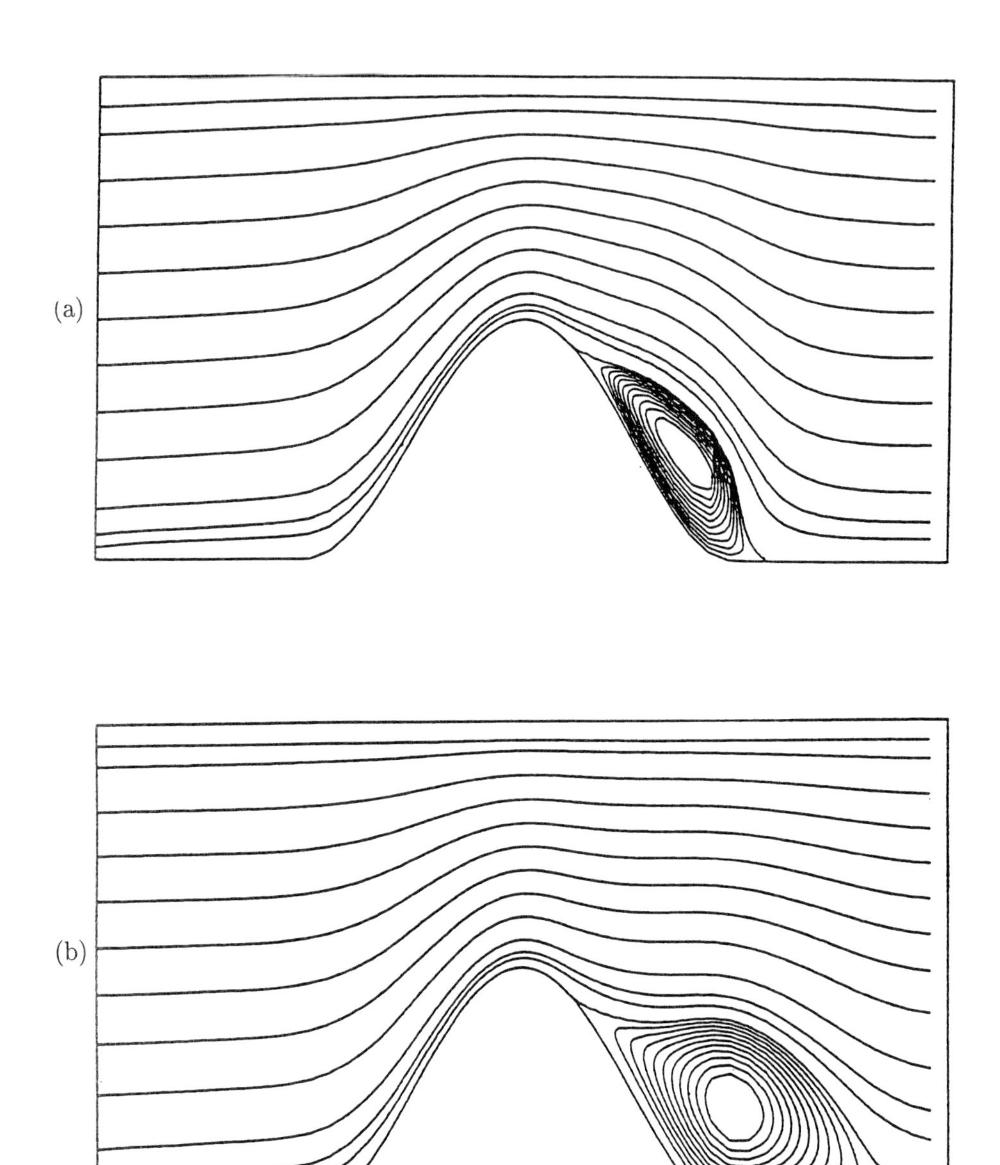

Figure 4. (a) Streamline contour for turbulent flow at Reynolds number 5000, and (b) Streamline contour for turbulent flow at Reynolds number 10000

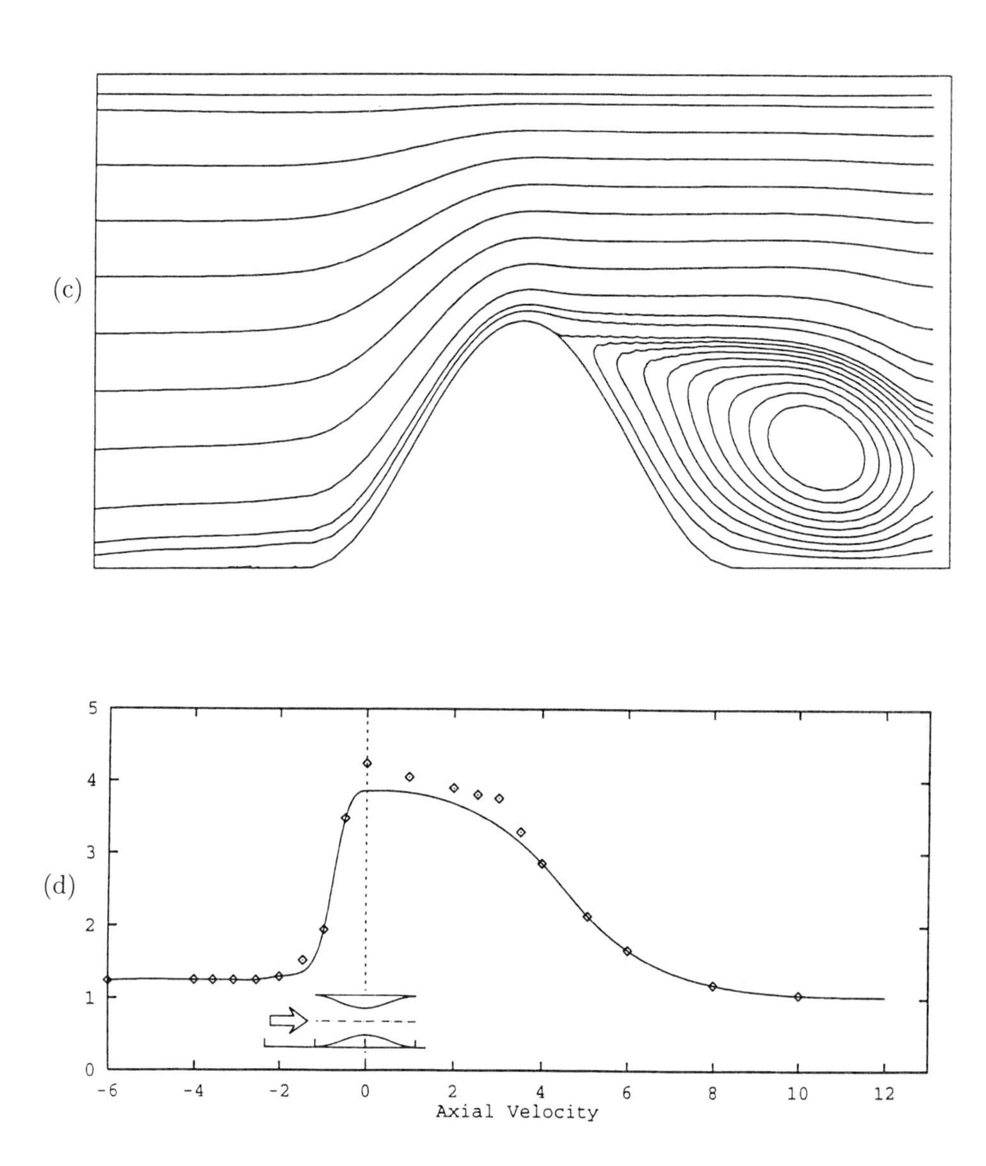

Figure 4. (c) Streamline contour for turbulent flow at Reynolds number 15000, and (d) Axial velocity variation along the axis dips at Reynolds number 15000. (◇) measurements, (—) computation)

5 Conclusions

Thus the present theoretical study of pulsatile flow in a stenotic tube confirms the view that the fluid dynamic characteristics of the flow are affected by the percentage of stenosis as well as the geometry of stenosis. The frequency of oscillation also influences the wall shear stress. However, a more realistic model for flow in stenotic tube of an arbitrary shape, for higher range of frequency parameter in the presence of turbulence relies on an accurate modelling of turbulence and sophisticated numerical integration of equations governing turbulent motion. Such a model has been developed, and currently being updated to incorporate a more accurate representation of turbulence.

Acknowledgements

This work has been dedicated to my dearest and loving brother, Shahram Sajjadi, for many years of moral support and the love and affection he has offered me so freely and unstintingly. I do greatly appreciate it.

Bibliography

1. Abramowitz, M., and Stegun, I.A. (1972). *Handbook of Mathematical Functions*, Dover.

2. Caro, C.G., and Saffman, P.G. (1965). Extensibility of blood vessels in isolated rabbit lungs. *J. Physiol.*, **178**, pp. 193–210.

3. Deshpande, M.D., and Giddens, D.P. (1980). Turbulence measurements in a constricted tube. *J. Fluid Mech.*, **97**, pp. 65–89.

4. Erdélyi, A. (Editor) (1954). *Higher Transcendental Functions* **III**, McGraw-Hill.

5. Fry, D.L. (1968). Acute vascular endothelial changes associated with increased blood velocity gradients. *Circulation Research*, **22**, p. 165.

6. Mehrotra, R., Jayaraman, G., and Padmanabhan, N. (1985). Pulsatile blood flow in a stenosed artery – a theoretical model. *Med. Biol. Eng. Comput.*, **23**, pp. 55–62.

7. Padmavathi, K., and Rao, R.A. (1992). A theoretical model of pulsatile flow in an arterial stenosis. *Physiological Fluid Dynamics III*, Editors: N.V.C. Swamy and M. Singh, Springer Verlag, pp. 144–148.

8. Sajjadi, S.G., and Aldridge, J.N. (1995). Prediction of turbulent flow over rough asymmetrical bed forms. *Journal of Applied Mathematical Modelling*, **19(3)**, p. 139.

9. Sajjadi, S.G., and Waywell, M.N. (1997). Application of roughness dependent boundary conditions to turbulent oscillatory flows. *Int. J. Heat and Fluid Flow*, **18**, p. 368.

10. Sajjadi, S.G., and Feng, Y. (1998). Pulsatile flow in an arterial stenosis; an asymptotic and computational study. Submitted for publication.

11. Wild, R., Pedley, T.J., and Riley, D.S. (1977). Viscous flow in collapsible tubes of slowly varying elliptical cross-section. *J. Fluid Mech.*, **81**, pp. 273–279.

Study of Newtonian and Non-Newtonian Fluid Flow in a Channel with a Moving Indentation

P. Neofitou, D. Drikakis[1] and M.A. Leschziner

Department of Mechanical Engineering, University of Manchester Institute of Science and Technology

Abstract

Numerical simulations of unsteady flow in a channel with a laterally moving indentation in one wall are presented. This paper focuses on the simulation of Newtonian and non-Newtonian flow in the above channel as part of a larger CFD research programme aimed at developing numerical and physical models for biological flows. The numerical results have been obtained by solving the Navier-Stokes equations in conjunction with the space-conservation law for a moving, compressing grid. Spatial resolution is achieved by a third-order discretisation scheme. The present results are in agreement with past Newtonian-flow computations reported in the literature. Comparisons between the Newtonian and non-Newtonian results reveal some differences in the flow patterns, especially regarding the formation of the vorticity waves downstream of the indentation.

1 Introduction

In the context of bio-fluid mechanics, the study of unsteady flow in channels with moving boundaries is largely motivated by the need to understand biological flows in collapsible tubes such as arteries and veins. Since *in vivo* arterial flows involve many physical issues which need to be modelled, simpler geometrical models are usually employed which, however, reproduce many of the important phenomena occuring in the real flow, such as unsteadiness, moving walls, separation and propagation of vorticity waves. Such a model is the flow in a channel with a moving indentation in one wall [1, 2]. Pedley and his co-workers have in the past studied, both experimentally and computationally, the flow downstream of a laterally moving indentation focusing especially on the generation and propagation of wave crests.

The above flow was considered in the present study as a basis for validating our numerical model which is based on the Navier-Stokes equations in generalised curvilinear co-ordinates in conjunction with the space conservation law for the case of moving, compressing grids. Moreover, the same case is used here

[1]Corresponding author, drikakis@umist.ac.uk.

to demonstrate the effects of non-Newtonian flow properties on the separated flow downstream of the moving indentation. This is done by employing a modified Casson model for the shear stress, as proposed by Papanastasiou [3]. The present results reveal that the above non-Newtonian model slightly supresses flow separation during the evolution of the unsteady flow.

2 Numerical modelling

2.1 Governing equations

The numerical model is based on the unsteady Navier-Stokes equations for an incompressible fluid. The coupling between the continuity and momentum equations is achieved via the artificial compressibility formulation, while the equations are employed for a curvilinear co-ordinate system and moving grids. The above system of equations is written as:

$$(JU)_\tau + (JU^*)_t + (E_I)_\xi + (G_I)_\zeta = (E_V)_\xi + (G_V)_\zeta. \tag{2.1}$$

The unknown solution vector U is:

$$U = (p/\beta, u, w)^T \tag{2.2}$$

where $U^* = (1, u, w)^T$, p is the pressure, u and w are the velocity components, and β is the artificial compressibility parameter. For the case of unsteady flows the parameter t denotes the real time, while τ is the pseudo-time in which sub-iterations are performed.
The inviscid fluxes E_I, G_I and the viscous fluxes E_V, G_V for a moving/compressing grid are written as:

$$\begin{aligned}
E_I &= J(U^*\xi_t + \tilde{E}_I\xi_x + \tilde{G}_I\xi_z)\\
G_I &= J(U^*\zeta_t + \tilde{E}_I\zeta_x + \tilde{G}_I\zeta_z)\\
E_V &= J(U^*\xi_t + \tilde{E}_V\xi_x + \tilde{G}_V\xi_z)\\
G_V &= J(U^*\zeta_t + \tilde{E}_V\zeta_x + \tilde{G}_V\zeta_z)
\end{aligned}$$

where the fluxes with "tildes" denote the corresponding Cartesian fluxes.

$$\tilde{E}_I = \begin{pmatrix} u \\ u^2 + p \\ uw \end{pmatrix}, \tilde{G}_I = \begin{pmatrix} w \\ uw \\ w^2 + p \end{pmatrix}.$$

$$\tilde{E}_V = \begin{pmatrix} 0 \\ \tau_{xx} \\ \tau_{xz} \end{pmatrix}, \tilde{G}_V = \begin{pmatrix} 0 \\ \tau_{zx} \\ \tau_{zz} \end{pmatrix}.$$

The terms τ_{ij} $(i, j = x, z)$ are the viscous stresses, and J is the Jacobian of the transformation from Cartesian to generalised co-ordinates. For the case of moving/compressing grids an additional equation, known as space-conservation law [4, 5], is also employed and solved in conjunction with the continuity and momentum equations:

$$J_t + (J\xi_t)_\xi + (J\zeta_t)_\zeta = 0 \tag{2.3}$$

2.2 Discretisation scheme

A characteristics-based method [6, 7] is used for the discretisation of the inviscid terms. A Riemann solution in each flow direction is formulated by splitting the inviscid equations into three one-dimensional equations. The primitive variables p, u and w at the cell faces of the computational volume are calculated as functions of their values on the characteristics. The latter are defined by the eigenvalues of the system of equations. The characteristic values are subsequently calculated by an upwind scheme up to fourth-order of accuracy. The viscous fluxes are discretised by second-order central differences, and the time integration is obtained by a fourth-order explicit Runge-Kutta scheme. Details regarding the method and the CFD code can be found in [7].

2.3 Non-Newtonian model

Modelling of the stress-deformation behaviour of blood is effected here via the Casson constitutive equation [8]. According to this relation, an unyielded structure is formed when the shear stress, τ_{ij}, falls below the yield stress, τ_y. Papanastasiou [3] proposed a modification of this equation by introducing a material parameter which controls the exponential growth of stress, thus avoiding the discontinuity inherent in the original viscoplastic model. In full tensorial form, the modifield relation is written as:

$$\overline{\overline{\tau}} = \left\{ \sqrt{\mu} + \sqrt{\frac{\tau_y}{|\dot{\gamma}|}} \left[1 - e^{-\sqrt{m|\dot{\gamma}|}} \right] \right\}^2 \overline{\overline{\dot{\gamma}}} \tag{2.4}$$

where $|\dot{\gamma}|$ is the magnitude of the rate-of-strain tensor

$$\overline{\overline{\dot{\gamma}}} = \nabla\vec{u} + \nabla\vec{u}^T, \tag{2.5}$$

and is given by

$$|\dot{\gamma}| = \left[\frac{1}{2} \left\{ \overline{\overline{\dot{\gamma}}} : \overline{\overline{\dot{\gamma}}} \right\} \right]^{1/2}. \tag{2.6}$$

The factor m is the stress growth exponent and $\dot{\gamma}$ is the shear rate. In this study m was set equal to 50. To track down the yielded/unyielded regions, the following criterion is employed:

$$\begin{aligned} \text{yielded:} \quad & |\tau| = \left[\tfrac{1}{2} \overline{\overline{\tau}} : \overline{\overline{\tau}} \right]^{1/2} > \tau_y \\ \text{unyielded:} \quad & |\tau| \leq \tau_y. \end{aligned} \tag{2.7}$$

According to the above relations, the material flows only when the magnitude of the extra shear-stress tensor exceeds the yield stress. Following Shah [9], the dynamic viscosity and yield stress for blood flow were set equal to $\mu = 0.00276 Pa \cdot s$ and $\tau_y = 0.01082 Pa$, respectively.

3 Results

The geometry of the channel with a moving indentation is shown in Figure 1. Numerical and experimental results for the Newtonian-flow case in this geometry and for various motions of the indentation have been discussed by Pedley and his co-workers, while in the present study the non-Newtonian flow effects are also investigated. The indentation starts flush (zero constriction), moves to a maximum indented position, according to the relation $H(t) = \frac{1}{2}(1 - cos 2\pi t)$ for $0 \leq t \leq 0.5$, and stays there ($H(t) = 1$ for $t > 0.5$). The flow has been studied for Reynolds and Strouhal numbers of $Re = 500$ and $St = 0.05$, respectively. Two computational grids were used, containing 64×34 and 121×34 grid points, respectively. Test computations showed that the coarser grid was inadequate for capturing all the wave crests appearing downstream of the indentation. The instantaneous streamline plots for the Newtonian case are shown in Figure 2. As previously shown by Ralph and Pedley [1], a finite train of vorticity waves is generated as the indentation moves into the channel. It can be clearly seen in Figure 2 that the wave crests continue to propagate downstream for some time after the indentation has stopped moving ($t = 0.5$ in Figure 2), while new eddies also develop. The formation of the wave patterns downstream of the indentation are in good agreement with the results of reference [1].

The Papanastasiou's model was used for a Casson number of 0.206. The latter is defined as $Ca = \frac{\tau_y D}{\mu V}$ where D is the channel height and V is a characteristic speed corresponding to $Re = 500$. The streamline patterns for the non-Newtonian case are shown in Figure 3 and indicate some differences in the generation and propagation of the wave crests downstream of the indentation. The non-Newtonian model makes the flow more viscous, resulting in a reduction of the eddies and, generally, a suppression of separation. A simulation was also performed for a Ca number equal to 0.784, and in that case it was found that the number of eddies was slightly reduced.

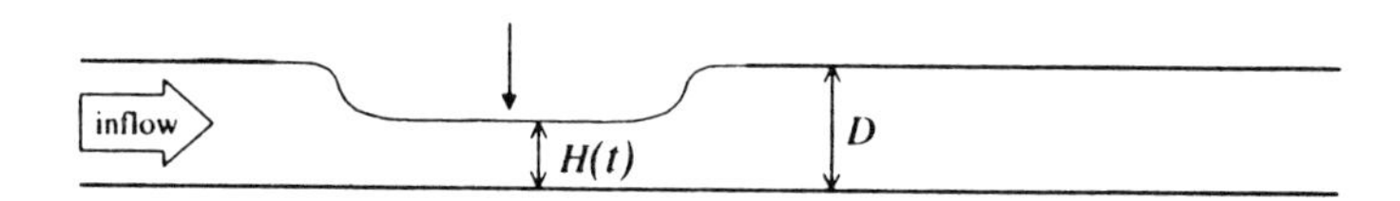

Figure 1. Schematic of the channel with a moving indentation in the upper wall

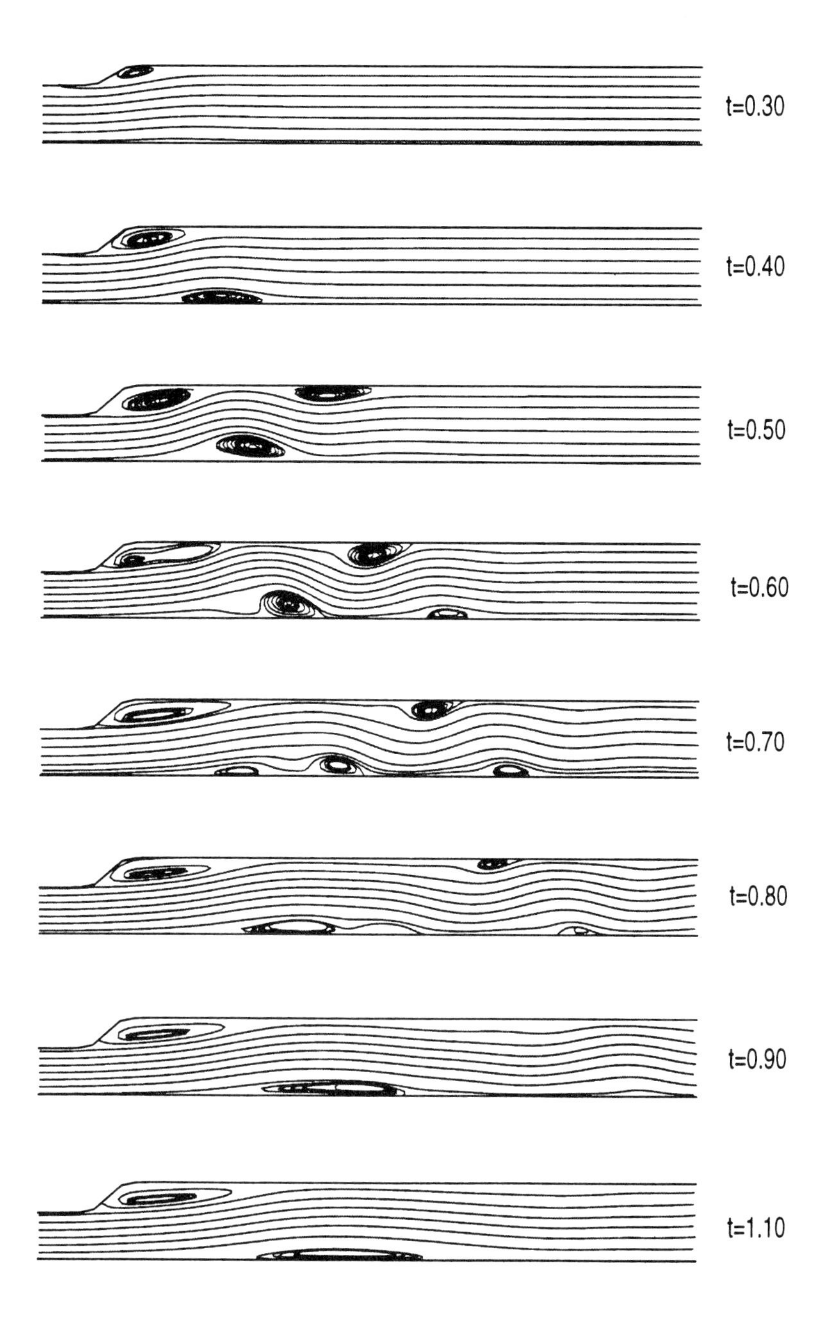

Figure 2. Streamlines plots at different time instants for the Newtonian flow case

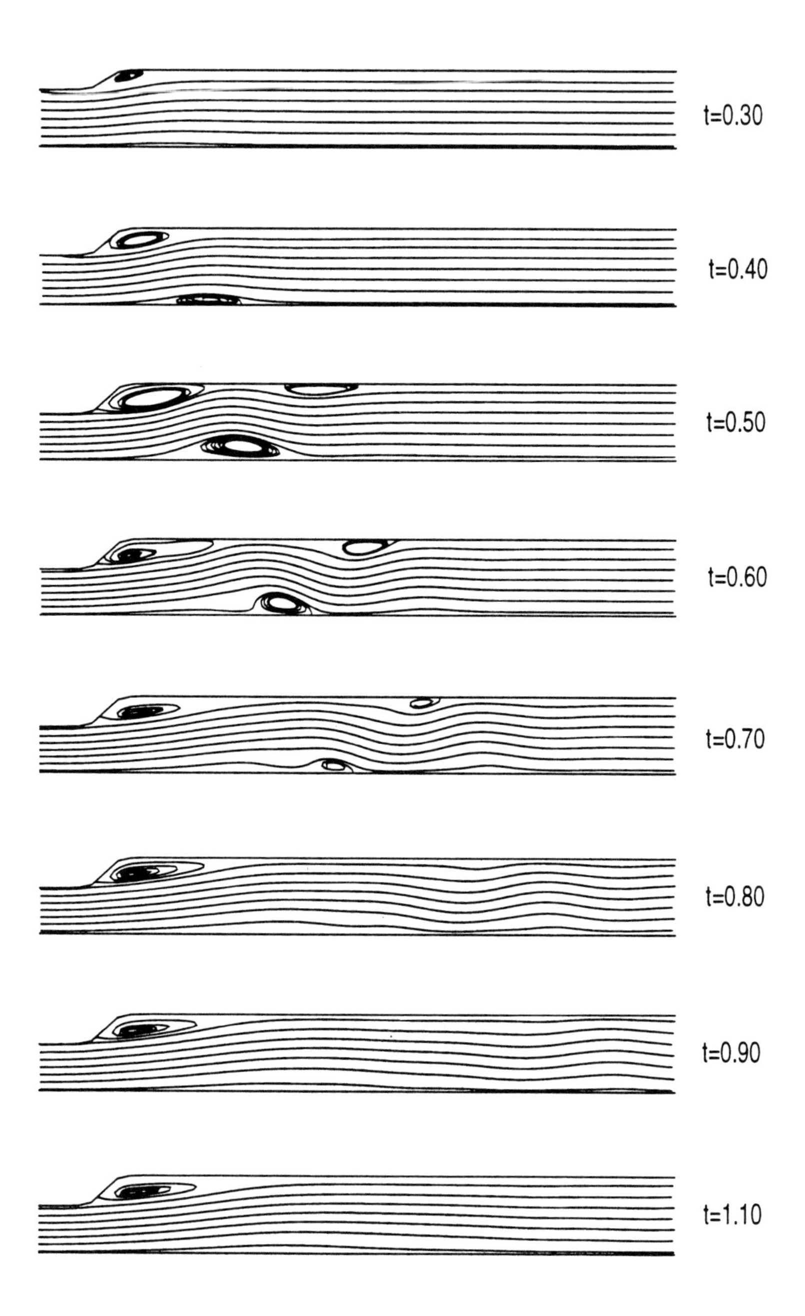

Figure 3. Streamlines plots at different time instants using the modified Casson model [3]

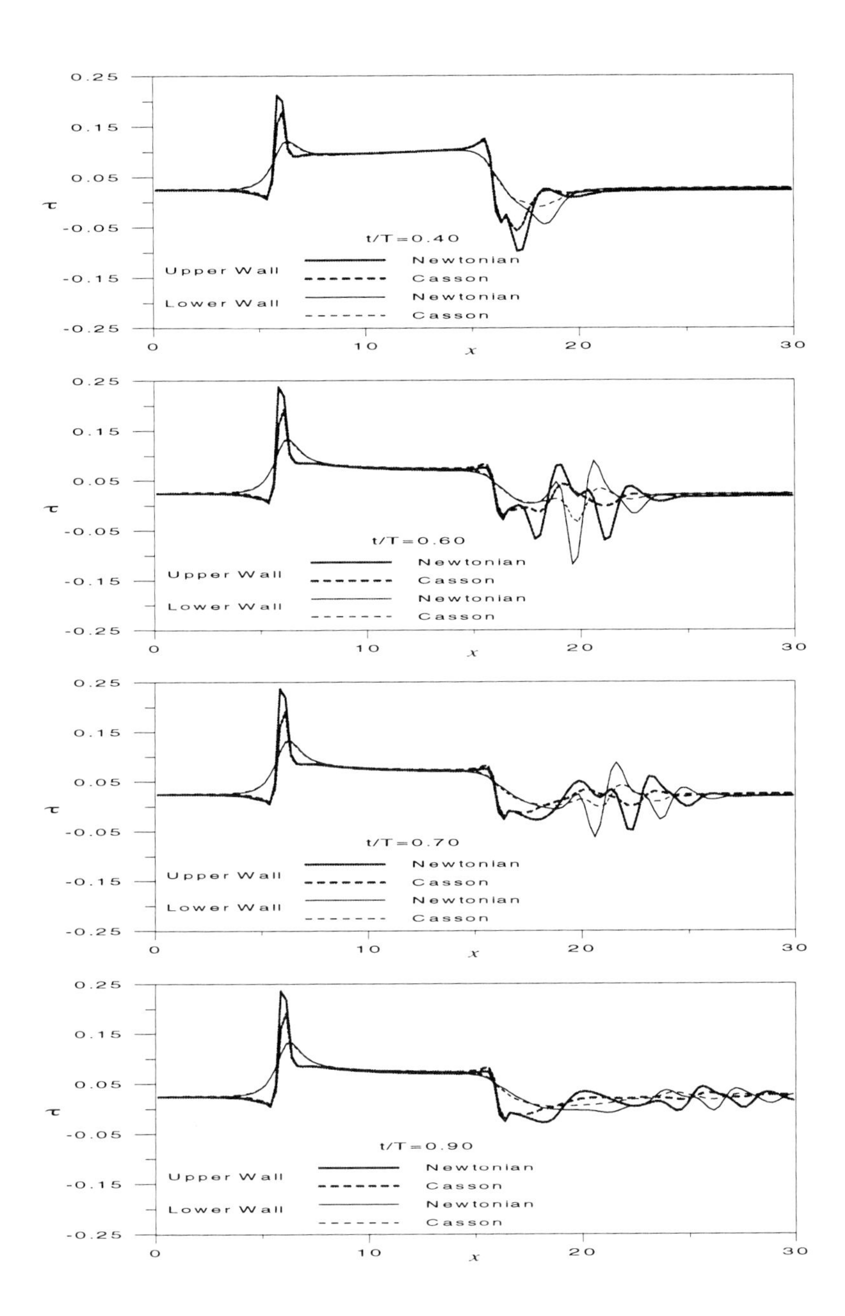

Figure 4. Wall-shear stress distributions at different time instants for the Newtonian and non-Newtonian flow cases

Finally, the shear-stress distributions on the lower and upper walls for the Newtonian and non-Newtonian case at different time instants are shown in Figure 4. The results reveal that the absolute wall-shear stress is reduced both on the upper and lower walls when the non-Newtonian model is employed. Moreover, due to the reduction of the wave crests in the non-Newtonian case, the fluctuations in the wall shear-stress downstream of the indentation are also reduced.

4 Conclusions

A study of Newtonian and non-Newtonian flow in a channel with a moving indentation in one wall was presented. The results for the Newtonian case were found in good agreement with corresponding results from the literature. A modified Casson model as proposed by Papanastasiou [3] was used for representing the non-Newtonian blood behaviour. The computations showed that this non-Newtonian model leads to a less separated flow in which some of the wave crests that existed in the Newtonian case are absent. Although the non-Newtonian results should be considered as preliminary, since there are no experimental data to validate the modified Casson model for this specific flow case, they qualitatively indicate that the non-Newtonian flow effects may be significant in flows in collapsible tubes. Therefore, further consideration to such phenomena needs to be given by employing and validating other rheological models in the case of blood flow.

Bibliography

1. Ralph, M.E. and Pedley, T.J. (1990). Flow in a channel with a time dependent indentation in one wall. *J. of Fluids Engnrg.*, *112*, 468–475.

2. Ralph, M.E. and Pedley, T.J. (1988). Flow in a channel with a moving indentation. *J. Fluid Mech.*, *190*, 87–112.

3. Papanastasiou, T.C. (1987). Flow of materials with yield. *J. Rheol.*, *31*, 385–404.

4. Trulio, J.G. and Trigger, K.R. (1961). Numerical solution of the one-dimensional hydrodynamic equations in an arbitrary time-dependent coordinate system. *Univ. of California Lawrence Radiation Laboratory Report UCLR-6522.*

5. Thomas, P.D. and Lombard, C.K. (1979). Geometric conservation law and its application to flow computations on moving grids. *AIAA J.*, *17*, 1030–1037.

6. Drikakis, D., Govatsos, P. and Papantonis, D. (1994). A characteristic based method for incompressible flows. *Int. J. Num. Meth. Fluids*, *19*, 667–685.

7. Drikakis, D. (1996). A parallel characteristcis-based method for 3D incompressible flows. *Advances in Engineering Software*, *26*, 111–119.

8. Casson, N. (1959). *Rheology of Disperse Systems*, Editor: C.C. Mill, Pergamon Press, New York.

9. Shah, H. (1980). Blood flow. *Advances in Transport Properties*, Volume I, Editors: A.S. Mujumdar and R.A. Mashelkar, Halsted Press, Wiley Eastern Limited, New Delhi.

Platelet Deposition in Stagnation Point Flow: An Analytical Study and Computational Simulation

T. David, S. Thomas and P.G. Walker

School of Mechanical Engineering, University of Leeds

Abstract

A mathematical and numerical model is developed for the adhesion of platelets in steady stagnation point flow. The model provides for a correct representation of the axi-symmetric flow and explicitly uses shear rate to characterise not only the convective transport but also the simple reaction mechanism used to model platelet adhesion at the wall surface. Excellent agreement exists between the analytical solution and that obtained by the numerical integration of the full Navier-Stokes equations and decoupled conservation of species equations. It has been shown that for a constant wall reaction rate modelling platelet adhesion the maximum platelet flux occurs at the stagnation point streamline. This is in direct contrast to that found in experiment where the maximum platelet deposition occurs at some distance downstream of the stagnation point. However, if the wall reaction rate is chosen to be linearly dependent on the wall shear stress then the results show that the maximum platelet flux occurs downstream of the stagnation point, providing a more realistic model of experimental evidence.

1 Introduction

The first stage of thrombogenesis is platelet adhesion. This may then be followed by platelet aggregation and the formation of mural thrombi [2] and [3]. Although the formation of thrombi in stasis is fairly well understood the influence of blood flow characteristics has yet to be fully investigated. Fluid dynamic studies of blood flow, in models of arteries, suggests a set of fluid dynamic conditions which appear to predispose thrombus formation, principally at arterial bifurcations [7], T-junctions and curved sections, in particular, Schoepheoster and Dewanjee [15] found that areas of flow stagnation or recirculation are shown to promote thrombus formation.

One of the areas where secondary flow occurs is downstream of a stenosis. Here a recirculation zone exists and further downstream, in the eventual transition back to parallel flow, a so-called "reattachment point" is formed [5,14,16,]. In this case a streamline directly impinges on a wall and contains a flow component analogous to that of stagnation point flow. Petschek et al. [10] suggested use of a stagnation point flow geometry for the investigation of thrombus growth under controlled conditions. Wurzinger et al. [20] suggested that there was a close correlation between flow conditions and the location of platelet deposits, in particular with stagnation point flow, where there are positive normal components of flow to the wall Tippe et al. [17] presented experiments in a stagnation point flow chamber whose flow was steady state. Their initial experiments showed that the number of platelets deposited at the lower wall surface was proportional to the number of platelets contacting the wall. In defining their experimental method it was shown that in the neighbourhood of the stagnation point the density of platelets was negligible but increased as the radial distance increased. Reininger et al. [11] and [12] used the same methodology to investigate the influence of flow on platelet adhesion to intact endothelium. Study of the published photographs indicated that there existed an area in the neighbourhood of the stagnation point which remained "platelet free", even though this area has the highest value for the normal component of velocity. The majority of platelets adhered in areas where the normal component was negligible. They continued to surmise that the platelet adhesion was, at least partially, mediated by the normal component of the convective particle transport of the stagnation point flow model even though the platelet adhesion seemed to be at a maximal density in a domain removed from the neighbourhood of the stagnation point, where flow convection towards the adhesion surface was, in contrast to the stagnation streamline, negligibly small.

Leonard et al. [8] stated that diffusion normal to the adhering surface was the determining factor in how blood components made contact with the adhering surface. Although convection was certainly important in bringing fresh reactants to the adhesion site. In particular they sited two important parameters which characterised the convection/diffusion equations. These were the Peclet number and a dimensionless surface reaction rate.

Turitto and Baumgartner [18] assumed that steady state deposition of platelets onto cylindrical tube surfaces was controlled by a diffusion process normal to the surface whose values changed as a function of shear rate. Their mathematical analysis made use of a transformation which included the assumption of a spatially independent linear velocity profile within the mass transfer boundary layer. They were able, by comparing with experimental data, to provide a power law relating diffusion coefficients to shear rate, however this was for whole blood where the presence of red blood cells is known to enhance the diffusivity of platelets.

Affeld et al. [1], using a stagnation point flow cell and platelet rich plasma (PRP), proposed that platelet deposition was maximal at the position of a "critical" shear rate. They hypothesised that the diffusion "collision theory" of platelet deposition was not a plausible one and that it seemed more likely that a critical shear rate was required to fully model the process where thrombin was convected downstream to activate further platelets. Pre-activated platelets were used in the experiment and hence the flux of thrombin emanating from the adhered platelets would seem to have little effect on already activated platelets.

Other effects apart from thrombin convection may provide the explanation for the spatial variation in platelet deposition.

Despite the considerable amount of published work on platelet adhesion there are still a considerable number of important unanswered questions regarding the modelling of platelet adhesion in flowing blood. Weiss [19] proposed that shear rate may have an effect on reaction rate especially for the case of the binding sequences between von Willebrand factor and GbIIb-IIIa. Hazel and Pedley [4] have also shown that pulsatile flow in the stagnation point region gives use to significant spatial variation in time averaged wall shear stress. In the light of this a novel model is developed below which incorporates convection, diffusion and reaction in a non-constant shear rate fluid flow.

2 Mathematical model

We begin by assuming a steady-state axi-symmetric jet centred at the origin O, where z denotes the co-ordinate normal to the surface whilst R is the radial position measured away from the origin. We assume that mass diffusion in the stream-wise radial (R) direction is negligible to that in the axial (z) direction and that the diffusion coefficient is spatially independent. The constant density conservation non-dimensional equation for the mass fraction of the ith species ϕ_i can now be written in the boundary layer type form of

$$u\frac{\partial \phi_i}{\partial R} + \nu\frac{\partial \phi_i}{\partial z} = \frac{1}{Pe_i}\frac{\partial^2 \phi_i}{\partial z^2} \qquad i = 1, \ldots, N \tag{2.1}$$

here ϕ_i is the ith species mass fraction, u and v are the velocity components for the R and z directions respectively. Pe, the Peclet number is defined as

$$Pe_i = \frac{UL}{D_i}$$

D_i is the Fickian diffusion coefficient and N denotes the total number of species participating within the domain. The diffusion coefficient is determined by the use of the Stokes-Einstein equation, giving a Peclet number of approximately $Pe = 6 \times 10^7$. For this case the velocity boundary layer profile is large compared to the species mass transfer boundary layer and can therefore be assumed to have a linear form. By using a von Mises transformation [13], which condenses the convective terms, and a similarity variable η similar to that derived by Kestin and Persen [6] given by

$$\eta = z\frac{\left\{\frac{\tau_w(R)}{\mu}\right\}^{\frac{1}{2}} Pe^{\frac{1}{3}}}{\left[9\int_0^R \left\{\frac{\tau_w(\gamma)}{\mu}\right\}^{\frac{1}{2}} d\gamma\right]^{\frac{1}{3}}} = z\beta(R), \tag{2.2}$$

where $\tau_w(R)$, the wall shear stress function whose values are known a priori, the species conservation equation becomes

$$\frac{d^2\phi_i}{d\eta^2} + 3\eta^2\frac{d\phi_i}{d\eta} = 0 \quad i = 1, \ldots, N. \tag{2.3}$$

For the case presented here it is assumed that no aggregation occurs in the bulk fluid (blood) and that the phenomenon of platelet adhesion is represented by a simple reaction boundary given by

$$\frac{\partial\phi_i}{\partial\eta}\frac{\partial\eta}{\partial z} = \kappa\phi_i \mid_{\text{wall}}$$

Weiss [19] postulated that the activation/adhesion of platelets is "controlled" somewhat by the local fluid shear stress. Using this as an extension to the model, the reaction rate at the cell chamber surface can be given a functional representation of the form

$$\kappa(R) = \kappa_0 + \alpha\tilde{\tau}(R) \tag{2.4}$$

where $\tilde{\tau}$ is non-dimensional shear stress at the wall evaluated from the numerical solution.

At some large distance from the reaction surface we have that

$$\eta \to \infty; \phi_{pl}(\eta) \to \phi_{pl\infty}. \tag{2.5}$$

It is easy to show that the solution to this problem is given by

$$\overline{\phi}_{pl}(\eta) = \frac{\phi_{pl}(\eta)}{\phi_{pl\infty}} = \frac{3}{\Gamma\left(\frac{1}{3},\infty\right)}\left(1 - \frac{3\beta(R)}{\kappa\Gamma\left(\frac{1}{3},\infty\right)+3\beta(R)}\right)\int_0^n e^{-\gamma^3}d\gamma$$
$$+\frac{3\beta(R)}{\kappa\Gamma\left(\frac{1}{3},\infty\right)+3\beta(R)}. \tag{2.6}$$

Here

$$\Gamma\left(\frac{1}{3},\infty\right) = \int_0^\infty e^{-t}t^{-2/3}dt.$$

The numerical model used the full steady-state non-dimensional momentum and conservation of species equations. Both momentum and species conservation equations were solved using the finite element commercial software code FIDAP. The domain was discretised into simple 8 noded quadralaterals, where the velocity was approximated by a quadratic formulation and the pressure determined through a piecewise continuous approximation. The form of $\beta(R)$ was found using the solution developed from the numerical model The non-dimensional shear stress at the wall was evaluated as a function of radial distance at a number of discrete points and interpolated using a polynomial format as shown below

$$\tilde{\tau}_w(R) = \sum_{k=1}^{12}\alpha_k R^k, \quad R \in [0,2]. \tag{2.7}$$

3 Results and Discussion

Figure 1 shows the normalised platelet concentration at the surface of the flow cell as a function of radial distance for both numerical and analytical models where κ, a constant reaction rate, takes various values. The agreement between these two models is excellent.

Figure 2 shows the corresponding platelet flux, q, to the surface for two constant reaction rates.

Figure 3 shows platelet concentration as a function of radial distance for shear rate dependent reaction rate at the flow chamber surface where κ_0 is non-zero ($\kappa_0 = 8951$). For κ_0 increasing the steady-state platelet concentration is reduced at $R = 0$.

Figure 4 shows non-dimensional platelet flux as a function of radial distance for various values of α with $\kappa_0 = 190$.

We note that $\beta(R)$ is a monotonic decreasing function over the domain of interest and we can easily show that, for a constant reaction rate, the platelet concentration and hence the platelet flux to the surface are also monotonic decreasing functions. Hence it can be seen that, for all values of the constant reaction rate κ presented, the platelet concentration and, more importantly, the flux is maximal at the stagnation point. The result of a maximum of platelet

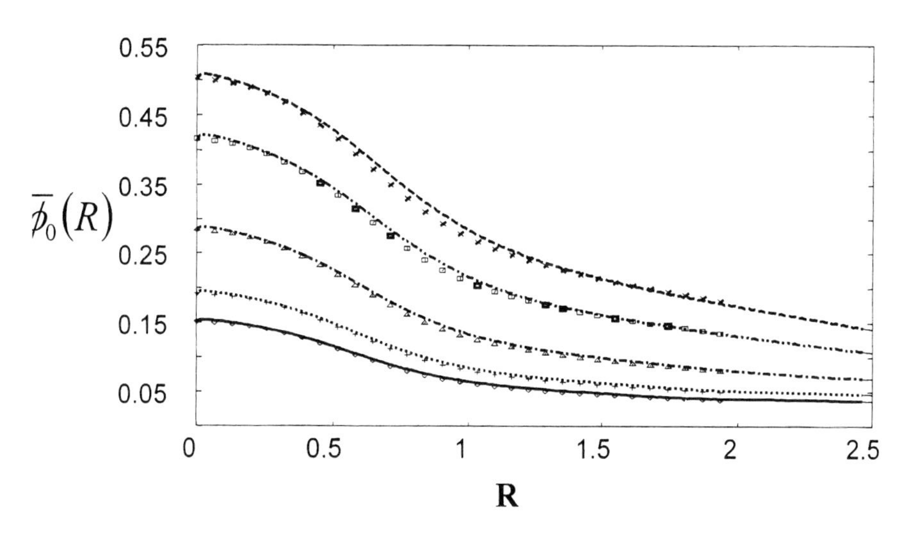

Figure 1. Normalised platelet concentration as a function of radial distance for constant reaction rate at the adhesion surface. Symbols are the analytic solution, lines are the numerical solution (◇ $\kappa_0 = 8951$; + $\kappa_0 = 5033$; - $\kappa_0 = 1812$; □ $\kappa_0 = 559$; x $\kappa_0 = 277$)

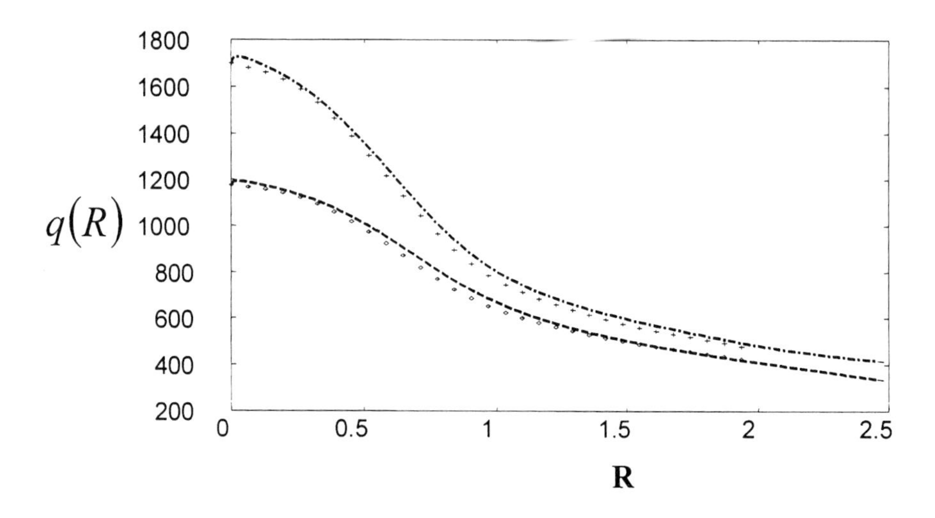

Figure 2. Non-dimensional platelet flux, q, as a function of radial distance at the flow chamber surface for two constant reaction rates. ($+\ \kappa_0 = 1812$; $.\diamondsuit\ \kappa_0 = 277$)

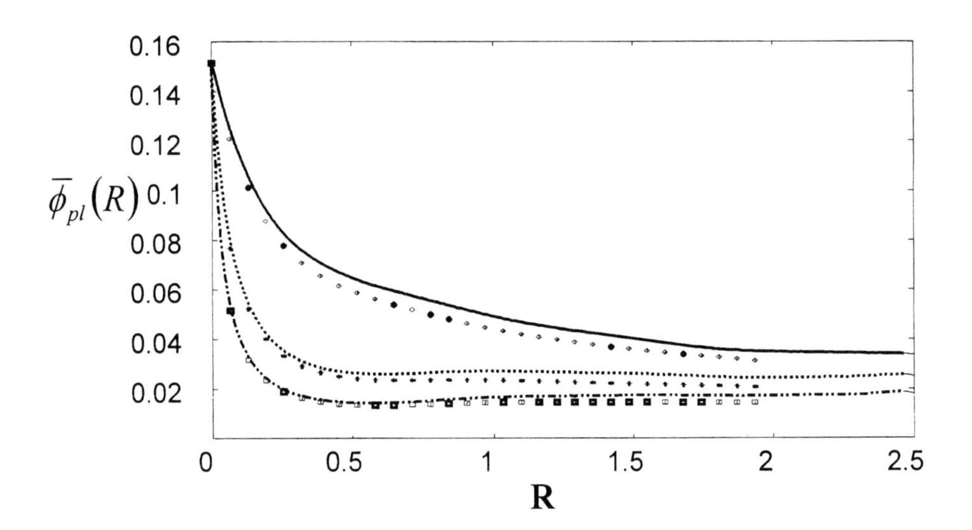

Figure 3. Platelet concentration as a function of radial distance for non constant reaction rate at the flow chamber surface where $\kappa_0 = 8951$

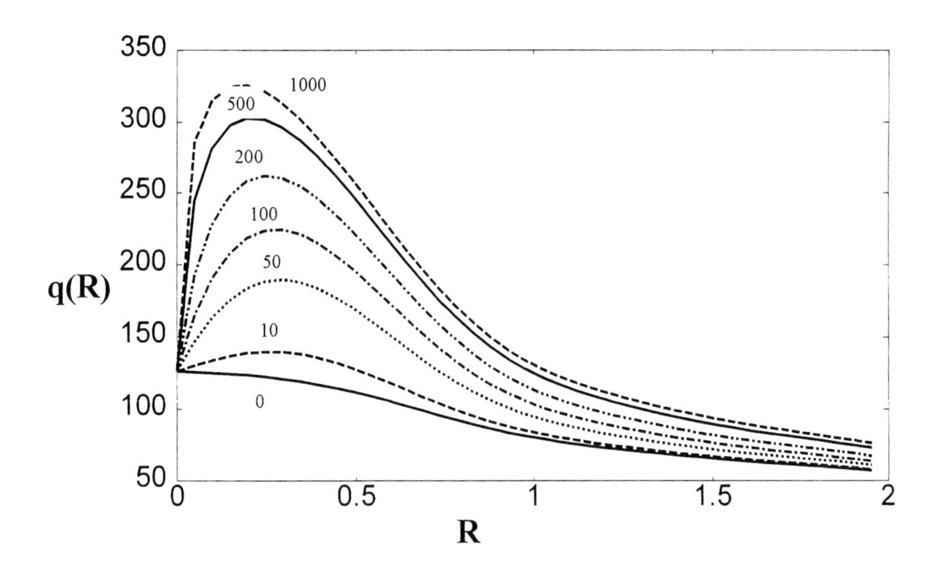

Figure 4. Platelet Flux, q, as a function of non-dimensional radial distance for various values of α (0,10,50,100,200,500,1000), with $\kappa_0 = 190$

flux at the stagnation point or high rate of adhesion at this point is in opposition with the results of platelet deposition experiments using stagnation point flow [1,20,21,22].

The occurrence of the flux maximum at the stagnation point leads to the important conclusion that, on comparing with experimental results of platelet adhesion in stagnation point flow, platelet adhesion at the lower wall of an experimental flow cell cannot be successfully modelled using a constant reaction rate function, even though both convection and diffusion is fully taken into account.

When the wall reaction rate is not constant then both concentration and flux profiles at the wall change. If we assume that the reaction varies as a linear function of shear rate as given by Equation (2.4). When the constant term κ_0 is zero then the platelet concentration is a maximum at the stagnation point. When κ_0 is non-zero then the platelet concentration at the stagnation point decreases as shown by Figure 3. The concentration profiles still have a minimum some distance from the stagnation point. This point of minimum concentration is to be determined as shown later in this section. The platelet flux profile for a shear rate dependent reaction rate is significantly different from the case of constant reaction rate. Figure 4 presents profiles for increasing values of the linear coefficient α ranging from 0 to 1000. This figure shows that the maximum platelet flux now occurs at some non-zero distance from the stagnation point.

Additionally as α increases the reaction rate increases and thus the position of maximum flux tends towards the stagnation point, $R = 0$.

We now look at the platelet flux with both constant and shear rate dependent reaction rate. The shear rate maximum occurs for $R = 0.4$ (evaluated by simple numerical root finding techniques). Figure 2 indicated that the flux for constant reaction occurred at the stagnation point. For this case it is easy to show that the maximum platelet flux is at $R = 0$, since by definition the flux is simply a multiple of the platelet concentration at the wall. For the non-constant reaction rate Figure 4 shows that the platelet flux maximum occurs at a value of R less than the maximum shear rate.

4 Conclusion

A mathematical model of platelet adhesion, using a high Peclet number approximation in a stagnation point flow chamber, has been presented which incorporates convection, surface reaction and diffusion. Excellent agreement exists between the analytical solution and that obtained by the numerical integration of the full Navier-Stokes equations and decoupled conservation of species equations.

It has been shown that for a constant wall reaction rate modelling platelet adhesion the maximum platelet flux occurs at the stagnation point streamline. This is in direct contrast to that found in experiment where the maximum platelet deposition occurs at some distance downstream of the stagnation point. However if the wall reaction rate is chosen to be linearly dependent on the wall shear stress then the analysis (and the numerical model) shows that the maximum platelet flux occurs downstream of the stagnation point, modelling experimental evidence. In summary

1. for constant wall reaction rate the maximum platelet flux and concentration occur at the stagnation point;

2. for a wall reaction rate, linearly dependent on wall shear stress, the platelet flux occurs at a radial position **less** than the position of maximum wall shear stress;

3. for a wall reaction rate, linearly dependent on wall shear stress, the platelet concentration occurs at a radial position **greater** than the position of maximum wall shear stress.

We note that the convection of thrombin is certainly important for the case of modelling the activation of platelets and their subsequent adhesion. However, once the platelets are activated and for situations where diffusion and reaction are not too dissimilar in magnitude, it seems likely that the model of platelet adhesion reaction becomes the important parameter in determining the flux distribution; especially if it varies as a function of wall shear stress. This also seems to have been hypothesised by Leonard et al. [8] based on experiments done by Monsler [9].

On the basis of the excellent agreement between numerical and analytical results the approximation of high Peclet number and the associated model is considered to be an excellent framework to investigate further reaction and species models in complex flows especially near reattachment points.

Bibliography

1. Affeld, K., Reininger, A.J., Gadischke, J., Grunert, K., Schmidt, S., and Thiele, F. (1995). Fluid mechanics of the stagnation point flow chamber and its platelet deposition. *Artificial Organs*, **19(7)**, pp. 597–602.

2. Friedman, L., and Leonard, E. (1971). Platelet adhesion to artificial surfaces: consequences of flow, exposure time, blood condition and surface nature. *Federation Proceedings*, **30(5)**, pp. 1641–1648.

3. Goldsmith, H. (1986). The microrheology of human blood. *Microvascular Research*, **31**, pp. 121–142.

4. Hazel, A., and Pedley, T.J. *Journal of Biomechanical Engineering.*

5. Karino, T., and Goldsmith, H. (1984). Role of blood cell–wall interactions in thrombogenesis and atherogenesis: a microrheological study. *Biorheology*, **21**, pp. 587–601.

6. Kestin, J., and Persen, L.N. (1962). The transfer of heat across a turbulent boundary layer at very high Prandtl numbers. *Int. J. Heat. Mass. Transfer*, **5**, pp. 355–371.

7. Kratzer, M., and Kinder, J. (1986). Streamline pattern and velocity components of flow in a model of a branching coronary vessel – possible functional implication for the development of localised platelet deposition in vitro. *Microvascular Research*, **31**, pp. 250–265.

8. Leonard, E.F., Grabowski, E.F., and Turitto, V.T. (1972). The role of convection and diffusion on platelet adhesion and aggregation. *Annals New York Academy of Sciences*, **201**, pp. 329–342.

9. Monsler, M., Morton, W., and Weiss, R. (1970). The fluid mechanics of thrombus formation. *AIAA Paper No. 70-787, AIAA 3rd Fluid and Plasma Dynamics Conf.*

10. Petschek, H., Adamis, D., and Kantrowitz, A.R. (1968). Stagnation flow thrombus formation. *Trans. Amer. Soc. Artif. Int. Organs*, **14**, pp. 256–260.

11. Reininger, A.J., Reininger, C.B., and Wurzinger, L.J. (1993). The influence of fluid dynamics upon adhesion of ADP stimulated human platelets to endothelial cells. *Thrombosis Research*, **71**, pp. 245–249.

12. Reininger, C.B., Graf, J., Reininger, A.J., Spannagl, M., Steckmeier, B., and Schweiberer, L. (1996). Increased platelet and coagulatory activity indicate ongoing thrombogenesis in peripheral arterial disease. *Thrombosis Research*, **82**, pp. 523–532.

13. Schlichting, H. (1960). Boundary layer theory. McGraw Hill Company Inc.

14. Schmid-Schonbein, H., and Wurzinger, L. (1986). Transport phenomena in pulsating post-stenotic vortex flow in arteries. *Nouvelle Revue Francaise d'Hematologie*, **28**, pp. 257–267.

15. Schoephoerster, R., and Dewanjee, M. (1996). Steady flow in an aneurysm model: correlation between fluid dynamics and blood platelet deposition. *Journal of Biomechanical Engineering*, **118**, pp. 280–286.

16. Schoephoerster, R., et al. (1993). Effects of geometry and fluid dynamics on regional platelet deposition on artificial surfaces. *Aterioslerosis and Thrombosis*, **13(12)**, pp. 1806–1813.

17. Tippe, A., Reininger, A., Reininger, C., and Reib, R. (1992). A method for quantitative determination of flow induced human platelet adhesion and aggregation. *Thrombosis Research*, **67**, pp. 407–418.

18. Turitto, V., and Baumgartner, H. (1975). Platelet deposition on subendothelium exposed to flowing blood: mathematical analysis of physical parameters. *Trans. Amer. Soc. Art. Int. Org.*, **21**, pp. 593–601.

19. Weiss, H.J. (1995). Flow-related platelet deposition on subendothelium. *Thrombosis and Haemostasis*, **74(1)**, pp. 117–122.

20. Wurzinger, L., Blasberg, P., van de Loecht, M., Suwelack, W., and Schmid-Schonbein, H. (1984). Model experiments on platelet adhesion in stagnation point flow. *Biorheology*, **21**, pp. 649–659.

21. Wurzinger, L., Blasberg, P., and Schmid-Schonbein, H. (1985). Towards a concept of thrombosis in accelerated flow: rheology, fluid dyanmics and biochemistry. *Biorheology*, **22**, pp. 437–449.

22. Wurzinger, L., Blasberg, P., Horii, F., and Schmid-Schonbein, H. (1986). A stagnation point flow technique to measure platelet adhesion onto polymer films from native blood. *Thrombosis Research*, **44**, pp. 401–406.

Fluid Dynamics in 3-D Models of Cavopulmonary Connection with an Extracardiac Lateral Conduit

F. Migliavacca*, M.R. de Leval*, G. Dubini, R. Pietrabissa† and R. Fumero†**

**Cardiothoracic Unit, Great Ormond Street Hospital for Children, London,*
***Department of Energetics, Politecnico di Milano, Milan, Italy, and*
†Department of Bioengineering, Politecnico di Milano, and CeBITeC, S. Raffaele Hospital, Milan, Italy

Abstract

Objective of this study is to quantify the haemodynamics in cavopulmonary connections with extracardiac lateral conduit. Four different models were studied with different length of inferior anastomosis (range 18–25 mm) and inclination of the conduit (33 and 47.5°). A commercial fluid dynamics code based on finite element method was utilised to compute Navier-Stokes and energy equations. Results showed that left-to right pulmonary flow ratio and percentage inferior caval blood to the left lung were the highest with the smallest anastomosis: 1.35 and 83.26%, respectively. Dissipated energy power percentage was the highest with the largest anastomosis than with the smallest (19.4% vs. 15.8%). An extracardiac lateral conduit when performing a total cavopulmonary connection

- diverts much flow to the left lung;
- shows higher energy losses if compared with a connection with intraatrial tunnel.

1 Introduction

In recent years refinements of the Fontan procedure have been widely described. Since the original aorto-pulmonary conduit performed for the first time by Fontan and Baudet [1], modifications have included the insertion of an intraatrial lateral tubular PTFE baffle [2] with or without fenestration and, at the beginning of the nineties, the construction of a new form of a right bypass: inferior vena cava-pulmonary artery extracardiac conduit [3].

Computational fluid dynamics (CFD) have been used to evaluate the haemodynamics of total cavopulmonary connection (TCPC) with an intraatrial tunnel and to refine the surgical procedure [4] and [5].

Basically, the TCPC consists in disconnecting the pulmonary artery from its ventricular origin and anastomosing the superior vena cava with the right

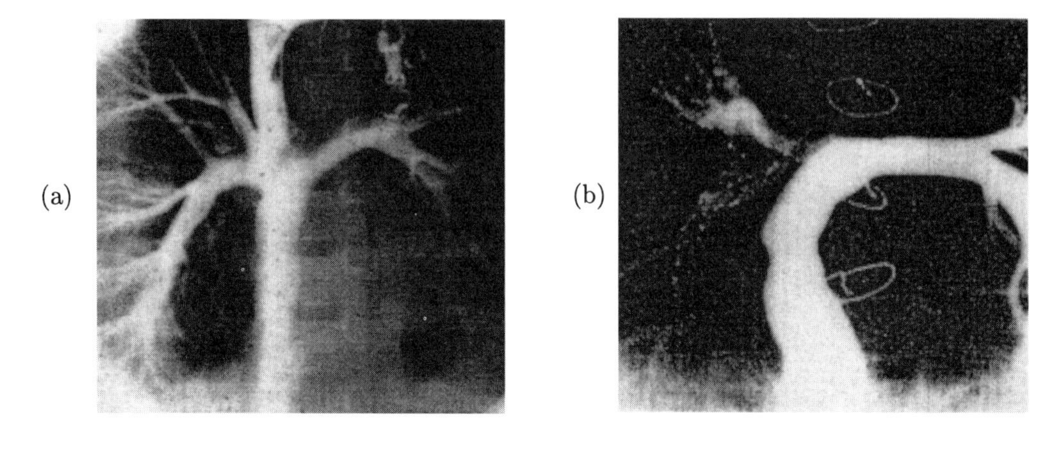

Figure 1. (a) TCPC with intraatrial tunnel; and (b) TCPC with extracardiac lateral conduit (from [6])

pulmonary artery. In the classical TCPC, the inferior vena caval blood is channelled through an intraatrial tunnel that is connected to the central and/or the proximal right pulmonary artery (Figure 1(a)).

Modifications of Fontan procedure have been proposed in recent years, among which the creation of a prosthetic extracardiac lateral conduit (ELC) [3] when performing the TCPC appears to have some advantages [6] over the procedure with an intraatrial tunnel or a right atrium-pulmonary anastomosis. In Figure 1(b) a typical postoperative angiograms (posteroanterior view) is presented [6].

Purpose of this study is to apply CFD to evaluate energy losses and lung blood flow distribution in models of TCPC in which an extracardiac lateral conduit is used in combination with a bidirectional Glenn shunt (the connection of the superior vena cava with the right pulmonary artery).

2 Materials and methods

Most of the methodologies applied in our previous works [4] and [5] are used in this study. We refer to those for a detailed description of the construction, of the assumptions and the limitations of the CFD models.

Each model of the TCPC with ELC consists of the superior vena cava (SVC), the extracardiac lateral conduit (ELC), and the left (LPA) and right (RPA) pulmonary arteries with their first bifurcations. The dimensions of the vessels have been kept as follow: SVC diameter equal to 12 mm, ELC diameter equal to 15 mm and RPA and LPA diameters equal to 12 mm. The investigated length of the ELC anastomosis (l_{ELC}) is 25 mm for model EX25, 22 mm for model

EX22 and 18 mm for model EX18. The distance between the centres of the SVC and the ELC anastomosis (offset) is equal to 0.3 cm for all the models. Angle between the ELC axis and RPA axis is equal to 33°. The fourth model has been constructed to assess the influence of inclination of ELC with respect to the pulmonary arteries (i.e. inclination of the insertion of ELC on the pulmonary arteries) on haemodynamics. Model EX18s has a length of the anastomosis equal to 18 mm as well as an inclination angle equal to 47.5^o. The afterload, represented by the lungs and the left atrium, has been taken into consideration by coupling the three-dimensional FEM model with a simple lumped parameter model with pulmonary arteriolar resistances (PAR) equal to 1.62 Woods unit (mmHg/l/min) and a left atrium pressure equal to 4 mmHg [5].

Two energetic indexes were calculated for the models:

- the hydraulic dissipated power W defined as follow:

$$W = \left(\frac{1}{2}\rho V_{\mathrm{ELC}}^2 + P_{\mathrm{ELC}}\right) Q_{\mathrm{ELC}} + \left(\frac{1}{2}\rho V_{\mathrm{SVC}}^2 + P_{\mathrm{SVC}}\right) Q_{\mathrm{SVC}}$$

$$-\sum_{i=1}^{4} \left(\frac{1}{2}\rho V_i^2 + P_i\right) Q_i \tag{2.1}$$

where P_i, Q_i and V_i represent the mean pressure, flow rate and velocity respectively on each of the four outlet sections of the pulmonary arteries and P_{ELC}, P_{SVC}, Q_{ELC}, Q_{SVC}, V_{ELC} and V_{SVC} the mean pressure, flow rate and velocity on each inlet section. ρ is the density of the blood and it has been assumed equal to 1,060 kg/m^3;

- the total energy loss coefficient C_e:

$$C_e = \frac{\left(\frac{1}{2}\rho V_{\mathrm{ELC}}^2 + P_{\mathrm{ELC}}\right) Q_{\mathrm{ELC}} + \left(\frac{1}{2}\rho V_{\mathrm{SVC}}^2 + P_{\mathrm{SVC}}\right) Q_{\mathrm{SVC}} - \sum_{i=1}^{4} \left(\frac{1}{2}\rho V_i^2 + P_i\right) Q_i}{\frac{1}{2}\rho \left(V_{\mathrm{ELC}}^2 Q_{\mathrm{ELC}} + V_{\mathrm{SVC}}^2 Q_{\mathrm{SVC}}\right)} \tag{2.2}$$

The latter index allows us to compare quantitatively the results of the different models, as the dimensions of the inlet diameters are the same for all the models and the inlet flow rates have been kept constant and equal to 1 l/min for the SVC and 2 l/min for the ELC, respectively. Reynolds numbers in the SVC and ELC are equal to 625 and 1,000, respectively.

In order to solve the Navier-Stokes and transport equations and to construct the four different models a commercial fluid dynamic code (FIDAP, Fluid Dynamics International, Evanston, Il, USA) based on the finite element method (FEM) has been utilised.

The repartition of ELC and SVC flows to the two lungs has been calculated with further simulations discarding the momentum equations and solving the following transport equation for one species (oxygen in the case):

$$\rho\left(\frac{\partial c}{\partial t} + \mathbf{u} \cdot \nabla c\right) = \rho \nabla \cdot (\alpha \nabla c). \tag{2.3}$$

The velocity fields ($\mathbf{u}$) previously calculated have been used so as to reduce the problem to an advection-diffusion one. Diffusivity (α) has been set equal to zero and concentrations (c) of the species at the inlet sections for ELC $[c_{\mathrm{ELC}}]$ and SVC $[c_{\mathrm{SVC}}]$ have been set equal to 100 and 0, respectively.

Repartition of ELC flow to the lungs has been calculated adopting the simple following relation:

$$Q_i \cdot [c_i] = Q_{\mathrm{ELC}-i} \cdot [c_{\mathrm{ELC}}] + Q_{\mathrm{SVC}-i} \cdot [c_{\mathrm{SVC}}] \tag{2.4}$$

where the index i refers to one of the outlet sections (RPA or LPA), and $Q_{\mathrm{ELC}-i}$ and $Q_{\mathrm{SVC}-i}$ are the portions of ELC and SVC flows directed towards the outlet i (RPA or LPA).

As $[c_{\mathrm{SVC}}]$ is equal to zero, it yields:

$$Q_{\mathrm{ELC}-i} = \frac{Q_i \cdot [c_i]}{[c_{\mathrm{ELC}}]}. \tag{2.5}$$

Then:

$$Q_{\mathrm{SVC}-i} = Q_i - Q_{\mathrm{ELC}-i}. \tag{2.6}$$

3 Results

Results of simulations are summarised in Table 1. If these results, in terms of energy losses and flow distribution to the lungs, are compared to the results of our previous study [4], it can be noted that the dissipated power and the total loss coefficients are generally higher. Furthermore, the blood flow is mainly directed to the left lung (Figure 2).

The velocity vector maps and the particle path plots in Figure 3 show that the ELC flow competes with the SVC flow.

Table 1. Results of the CFD simulations of ELC

Model	l_{ELC} [mm]	RPA-ELC angle [deg]	W [%]	C_e [-]	$Q_{\mathrm{LPA}}/Q_{\mathrm{RPA}}$ [-]	$Q_{\mathrm{ELC\text{-}LPA}}$ [%]	$Q_{\mathrm{ELC\text{-}RPA}}$ [%]	$Q_{\mathrm{SVC\text{-}LPA}}$ [%]	$Q_{\mathrm{SVC\text{-}RPA}}$ [%]
EX18	18	33.5	16.6	6.2	1.35	83.26	16.74	5.68	94.32
EX22	22	33.5	17.9	6.8	1.29	60.90	39.10	47.28	52.72
EX25	25	33.5	19.4	7.5	1.30	55.10	44.90	59.62	40.38
EX18s	18	47.5	15.8	5.8	1.20	59.06	40.94	45.33	54.67

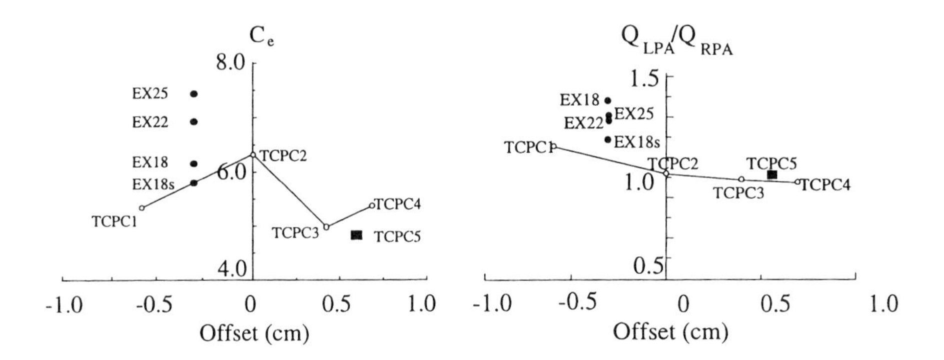

Figure 2. C_e and left to right flow ratio between the lungs in the four models studied. Results are compared to those from simulation of TCPC with intraatrial tunnel [4]

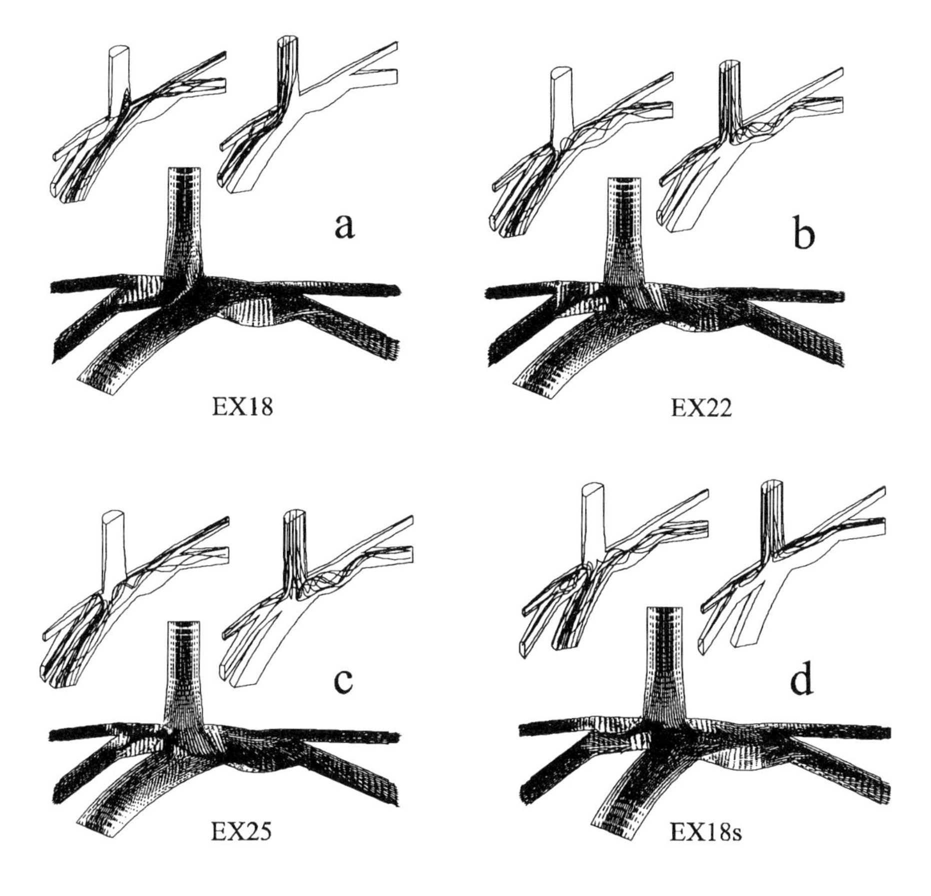

Figure 3. Velocity vector maps and particle path plots for the four models of TCPC with ELC

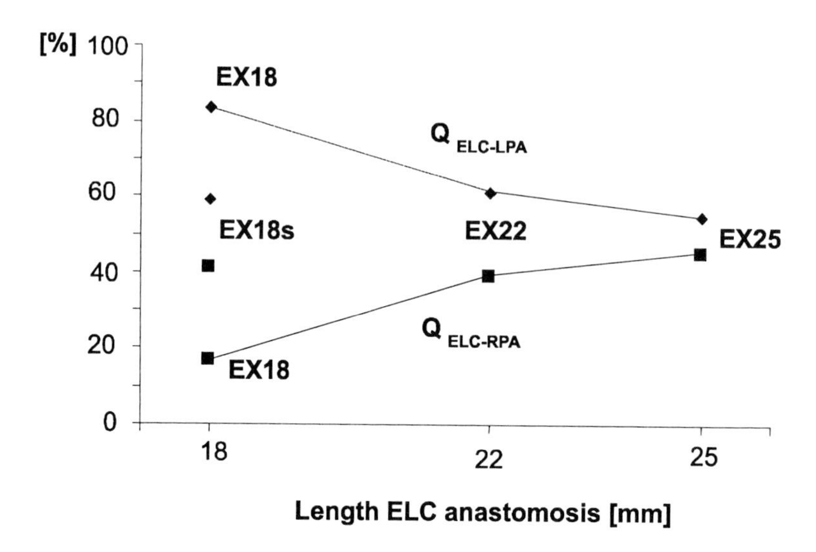

Figure 4. Percentage ELC caval blood flow distribution to the right ($Q_{\text{ELC-RPA}}$) and left ($Q_{\text{ELC-LPA}}$) lung

When increasing the length of the ELC anastomosis, SVC flow moves down towards the ELC. Each lung is perfused from both caval flows in models EX22 and EX25. The smallest anastomosis length (EX18) causes the LPA to be mainly perfused by the ELC blood (Figure 3(a)). The increase in the ELC inclination (model EX18s) leads to energy losses similar to the TCPC models with intraatrial tunnel (Figure 2), with the best lung flow distribution in the ELC model set. Moreover (particle path plot of Figure 3(d)), the bigger RPA-ELC angle inclination leads the ELC flow to split towards both lungs and it confines the SVC flow to the superior aspects of the pulmonary arteries.

Inclination of the ELC towards the left lung results in streaming of inferior vena caval flow to the left lung and therefore it may create a non-physiological distribution of hepatic venous blood to each lung. Figure 4 shows the contribution of caval flows to each pulmonary branch. It can be argued that a narrower anastomosis (model EX18) directs most of the inferior blood to the left lung. When enlarging the ELC anastomosis and increasing the RPA-ELC inclination, the flow from the lower part of the body is better split into the lungs.

4 Discussion

In this study we have focused the attention only on the haemodynamics of ELC in cavopulmonary connections, evaluated by means of computational fluid dynamics. In particular, we have calculated energy losses, left-to-right pulmonary flow ratio and inferior vena cava blood distribution to the lungs for four different configurations of ELC. Indeed, the knowledge of the amount of inferior venous flow directed to each lung seems to be important in the development of pulmonary arteriovenous malformations [7]. In particular these data may have a clinical significance with regard to the post-operative pleural effusion and the late development of pulmonary arteriovenous fistulae in the right lung, which does not receive hepatic blood. However, a discussion on advantages and disadvantages of the different techniques for the Fontan procedure is not the aim of this work.

In recent literature there have been several studies concerning the haemodynamics of the Fontan operation based on in vitro [8] and [9] or computational fluid dynamics models [4,5,10], but none of them has considered the presence of an ELC.

Results of this study have shown that an ELC, when performing the TCPC, diverts most of the inferior caval flow to the left lung. In the presence of a negative offset (i.e. SVC anastomosed to the RPA and ELC to the LPA) a bigger RPA-ELC inclination angle leads to a better left-to-right flow ratio and to minor energy losses. Enlargement of the ELC anastomosis redirects part of the inferior blood to the right lung as well.

Acknowledgements

This work has been supported by the British Heart Foundation.

Bibliography

1. Fontan, F., and Baudet, E. (1971). Surgical repair of tricuspid atresia. *Thorax*, **26**, pp. 240–248.
2. de Leval, M.R., Kilner, P.J., Gewillig, M., and Bull, C. (1988). Total cavopulmonary connection. A logical alternative to atriopulmonary connection for complex Fontan operation. *J. Thorac. Cardiovasc. Surg.*, **96**, pp. 682–695.
3. Marcelletti, C., Corno, A., Giannico, S., and Marino, B. (1990). Inferior vena cava-pulmonary artery extracardiac conduit: a new form of right heart bypass. *J. Thorac. Cardiovasc. Surg.*, **100**, pp. 228–232.
4. de Leval, M.R., Dubini, G., Migliavacca, F., Jalali, H., Camporini, G., Redington, A., and Pietrabissa, R. (1996). Use of computational fluid dynamics in the design of surgical procedures: application to the study of competitive flows in cavo-pulmonary connections. *J. Thorac. Cardiovasc. Surg.*, **111**, pp. 502–513.

5. Dubini, G., de Leval, M.R., Pietrabissa, R., Montevecchi, F.M., and Fumero, R. (1996). A numerical fluid mechanical study of repaired congenital heart defects. Application to the total cavopulmonary connection. *J. Biomechanics*, **29**, pp. 111–121. [Erratum. *J. Biomechanics*, (1996), **29**, p. 839.]

6. Laschinger, J.C., Redmond, J.M., Cameron, D.E., Kan, J.S., and Ringel, R.E. (1996). Intermediate results of the extracardiac Fontan Procedure. *Ann. Thorac. Surg.*, **62**, pp. 1261–1267.

7. Kawashima, Y. (1997). Cavopulmonary shunt and pulmonary arteriovenous malformations. *Ann. Thorac. Surg.*, **63**, pp. 930–932.

8. Sharma, S., Goudy, S., Walker, P., Panchal, S., Ensley, A., Kanter, K., Tam, V., Fyfe, D., and Yoganathan, A.P. (1996). In vitro flow experiments for determination of optimal geometry of total cavopulmonary connection for surgical repair of children with functional single ventricle. *J. Am. Coll. Cardiol.*, **27**, pp. 1264–1269.

9. Kim, Y.H., Walker, P.G., Fontaine, A.A., Panchal, S., Ensley, A.E., Oshinski, J., Sharma, S., Ha, B., Lucas, C.L., and Yoganathan, A.P. (1995). Hemodynamics of the Fontan connection: an in-vitro study. *J. Biomech. Eng.*, **117**, pp. 423–428.

10. Van Haesdonck, J.M., Mertens, L., Sizaire, R., Montas, G., Purnode, B., Daenen, W., Crochet, M., and Gewillig, M. (1995). Comparison by computerized numeric modeling of energy losses in different Fontan connections. *Circulation*, **92** (Supplement), pp. II322–II326.

Modelling of Flow Transport in Arteries

D.J. Doorly

Department of Aeronautics, Imperial College, London

Abstract

Computational procedures to model transport processes in large arteries are described. The procedures are based on hybrid Eulerian-Lagrangian techniques, in which the velocity field represented on a fixed Eulerian mesh is used to compute the trajectories of marker particles. Applications to the modelling of high Péclet number convection-diffusion and shear-induced damage are described. Extension of the procedure to enable numerical computation of magnetic resonance images of flow in arteries is briefly outlined.

1 Introduction

The cardiovascular system has evolved to transport the essential requirements for life (food, oxygen, heat and waste) rapidly to each living cell in man. Small organisms have no requirement for such a complex network of vessels, since over short distances diffusion alone is an effective transport mechanism. The time t taken for diffusion to transport a given species a distance x increases according to $t \propto \frac{x^2}{2D}$, where D is the coefficient of diffusion of the species in the medium; consequently the effectiveness of diffusion as a transport mechanism rapidly decreases with distance. By using a suspension of particles in a fluid (i.e. blood) which is forced around the body by the heart, convection becomes responsible for transport over large distances thus overcoming the limitations of diffusion. Although convection is the dominant means of transport within the large arteries, even here diffusion plays an important role, particularly in the flow near the vessel walls.

Blood is a complex fluid, comprising a suspension of particles (the formed elements: erythrocytes, leukocytes and platelets) in plasma. Detailed modelling of the dynamics of the particles in the flow will not however be considered here. Instead the particles will be idealised as being of infinitesimal size and of neutral buoyancy. We do not consider turbulent flows since for the most part the flow in arteries remains laminar, although it may be highly disturbed. The purpose of this work is to outline particle and particle-in-cell based approaches suitable for the modelling of the transport processes in arteries. There are several advantages of such techniques. Included among these are:

- they conveniently allow particle history effects to be incorporated (so that accumulated shear damage may be modelled),
- they provide a robust method of handling numerical difficulties which occur in convection-dominated transport, as in arterial flows,
- they provide a convenient basis for post-processing techniques, such as the simulation of flow measurement by magnetic resonance imaging,
- they provide a starting point for more exact particle modelling.

Before describing the numerical approach, diffusion is briefly discussed.

2 Diffusion coefficients

The diffusion coefficient for a solute in a liquid may be estimated from the Stokes-Einstein equation

$$D = \frac{k_B T}{6\pi\mu R_o} \tag{2.1}$$

where k_B, the Boltzmann constant, is 1.38 $\times 10^{-16} g \cdot cm^2/\mathrm{sec}^2 \cdot K$, T is the absolute temperature, μ is the fluid viscosity and R_o is the assumed spherical radius of the solute. This relation is only approximate, and there are various forms of correction for solute concentration, molecular shape etc.

For example, as described in [1], using (2.1) to estimate the diffusion coefficient D of oxygen in water at $25^o C$, (where the molecular radius of oxygen is assumed to be half its collision diameter as a gas, namely $1.73 \times 10^{-8} cm$), D is predicted to be $1.3 \times 10^{-5} cm^2/sec$, whereas the measured value for low concentration is $1.8 \times 10^{-5} cm^2/sec$ [1]. However (2.1) is nevertheless useful to provide a reasonable estimate, and indicates how the diffusion coefficients of larger molecules reduce with increasing size: D for species such as ADP or thromboxane A_2 is of the order of 2×10^{-6}, for thrombin it is approximately 4×10^{-7} and for fibrinogen $2.0 \times 10^{-7} cm^2/sec$.

The diffusion coefficient for a particle such as a platelet (of order $2\mu m$ diameter) would, if solely due to Brownian motion, be extremely small. Platelet diffusion in whole blood is greatly augmented due to interactions with the larger particles, which are predominantly erythrocytes. A value for platelet diffusion in whole blood of $10^{-7} cm^2/sec$ is often used [2]. Flow shear enhances the diffusion of particles as it increases their rotation, but it also has the effect of promoting a tendency towards seggregation of particles according to size. Models to account for the "drift" of platelets towards the wall, as the larger particles migrate from the highest shear regions have also been employed [3] and [4]. More detailed modelling of particles would consider Basset and Faxen forces [5], but we will neglect these in this review.

3 Transport by convection and diffusion

The equations for the transport of a passive scalar may be written in terms of the concentration c for a Newtonian incompressible fluid as:

$$\frac{\partial c}{\partial t} + \vec{u} \cdot \nabla c = \frac{1}{ReSc} \nabla^2 c = \frac{1}{Pe} \nabla^2 c \tag{3.1}$$

where the Schmidt number $Sc = \frac{\nu}{D}$, and Pe is the Péclet number. Numerical solution of this equation using Eulerian methods (where the concentration field is represented by values at the nodes of a fixed mesh) must overcome stability problems at high Péclet numbers, but this is not an issue with Lagrangian particle methods. In the Lagrangian approach the concentration field is carried by marker particles and the velocity of each particle is generally interpolated from the results of a computation on an Eulerian mesh. The transport equation is usually split into convective and diffusive operators, where the former is solved simply by integrating the particle velocity over a time step (using for example a Runge Kutta method) to obtain the convective displacement of each particle. The evolution of the entire concentration field follows from the motion of the entire particle assembly.

There are several means of treating the diffusion operator. Many of these have been developed for the solution of the vorticity transport equation, such as the random vortex method, [6]. The diffusion operator

$$\frac{\partial c}{\partial t} = \frac{1}{Pe} \nabla^2 c \tag{3.2}$$

can be interpreted in terms of the stochastic differential equation

$$dx = \sqrt{2/Pe}(dW_t) \tag{3.3}$$

where W_t is a Wiener process. If at every time step Δt the position coordinates of each of the particles is advanced from time step n to $n+1$ according to

$$x_k{}^{n+1} = x_k{}^{n} + \eta_k \tag{3.4}$$

where the η_k are Gaussian random variables $N(0, \sigma^2)$ with $\sigma^2 = 2\Delta t/Pe$, the effect of diffusion may be incorporated as an ensemble of realisations.

Alternatively, hybrid Eulerian-Lagrangian particle-in-cell methods may be used to solve for diffusive transport. In this approach the particle "strengths" are projected onto a fixed Eulerian grid. The diffusion operator is applied on the grid, and changes in grid values are reprojected back to the particles, with new particles being created as required. Volume weighted interpolation as shown in Figure 1(a) is used in this work to interpolate velocity fields, and to project from particles. The only difference is that we employ tetrahedral elements instead of the simple cuboid element-type shown. The particle-in-cell method has been developed for a number of applications, including solving for vorticity transport [7]. At high Péclet numbers, the changes in mesh quantities at each step may be very low so that separate particles to carry the change in mesh quantities may be required.

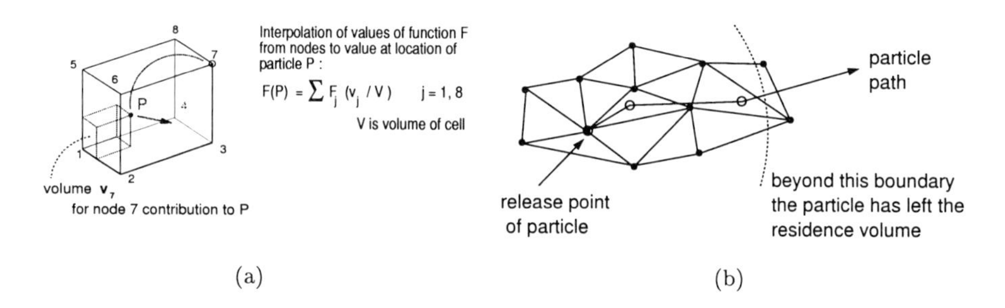

(a) (b)

Figure 1. (a) Volume weighted interpolation of nodal values (shown for simplest element), and (b) Residence time measure for unsteady flows

4 Model flow problem

The causes of arterial bypass graft failure are thought to be linked to adverse flow conditions, particularly in the vicinity of the anastomotic junctions. As a model problem, the flow entering a cylindrical artery from an equi-diameter cylindrical graft in a so-called "end-to-side configuration" is considered. The angle of the graft to the model artery is 45 degrees, and for most of the results we show here, all vessels are assumed coplanar. The portion of the artery upstream of the graft is assumed fully occluded.

4.1 Measures

Computational modelling may be used to recover quantities difficult to determine experimentally, for example residence times, and wall shear stresses. We would also like to examine the origin of particles which end up in long residence time regions, and to predict the integrated loading on particles through the stresses induced on them by the flow. The computational technique we use to predict the flowfield is the unstructured high order h/p finite element method of Sherwin et al. [8].

4.1.1 Residence time & wall shear

Given that platelets are believed to react rapidly, the occurence and location of quasi-transient zones of long residence time in unsteady flows is of particular interest. Residence time may be defined in terms of the delay before a particle escapes from a certain neighbourhood of a release point. Figure 1(b) illustrates this where an exaggeratedly large escape radius is shown for clarity.

The distribution of wall shear stress for the model artery is shown in Figure 2, together with the corresponding distribution for a "non-planar" type of graft configuration. The effects of non-planarity of the geometry of the vessels on the flow are striking, for example the shear stress is rendered considerably more

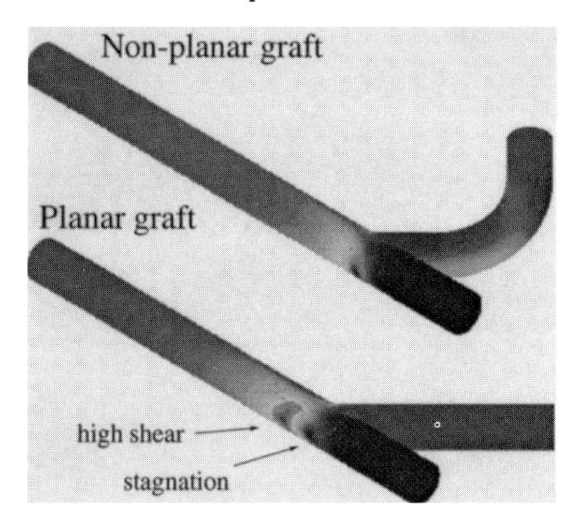

Figure 2. Comparison of shear stress at the distal anastomosis for idealized grafts, for steady flow at Re 250. Planar graft shows high shear stress peak downstream of planar-symmetrical stagnation point

uniform; this issue is considered in [9], [10] and [11] and will not be considered further here.

4.1.2 Particle paths and migration of particles

Computational aspects of interpolating particle velocities from a high order element to an individual particle will be dealt with elsewhere [12]. Results of particle tracking include simple colour tagging of particles according to their origin; for example in Figure 3 particles are coloured at inflow to the graft according to their radial distance from the origin under steady laminar flow at $Re_D = 290$. In a grey scale representation, the centre of the graft (dark blue) appears dark, as does the edge layer (red), whilst in between the particles appear lighter. Given that platelets for example may be activated by contact or near contact with thrombogenic surfaces, and considering the possible presence of activating surfaces such as stitches or trauma sites, we would like to understand where such particles end up. As the flow passes from graft to artery it is forced to curve, producing a Dean-type flow which leads to eruption of the near wall layer of flow affecting particle transport as shown in Figure 3. At the 2D location there is a migration of originally near-wall particles into the core of the flow, whereas on the opposite outer wall, particles originally near the graft centreline predominate.

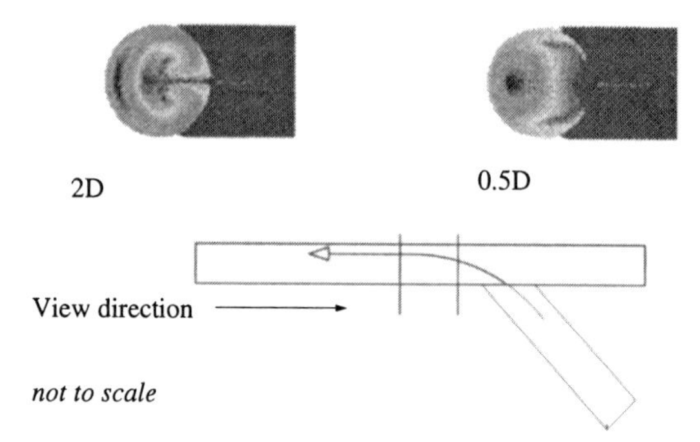

Figure 3. Distribution of particles at two sections of ideal model artery just downstream of anastomosis. Particles are coded according to radial position at inflow to graft. Just downstream of toe (0.5 diameters), particles near the centreline approach bed of artery. At 2 D downstream, eruption of the boundary layers as a Dean-type flow develops transports a "mushroom" shaped plume of particles from near wall regions

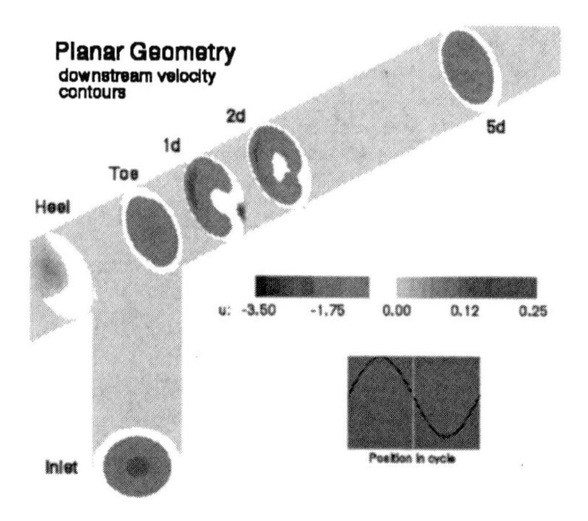

Figure 4. Axial velocity contours for unsteady flow in a planar graft at mean Re of 250. Low positive velocities are not shown, to allow recirculation regions on "bed" of model artery, i.e. opposite graft, and near "toe" to be visible. The recirculation at the toe is quite prominent, but that on the bed is confined to a thin layer

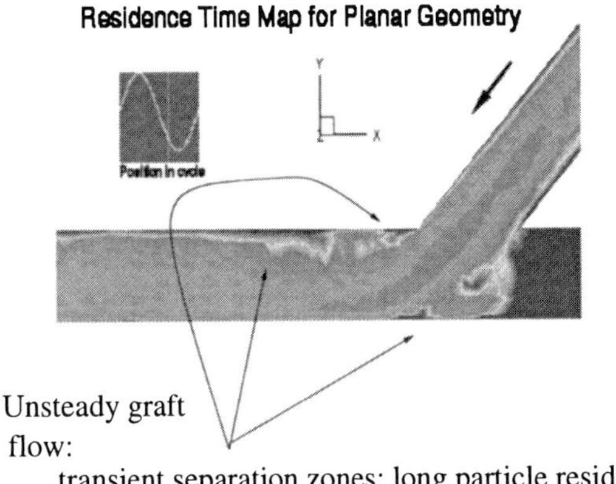

Figure 5. Transient long residence times (dark grey/red) near wall, in regions of separation, and throughout occluded end for planar graft (centre line section)

4.2 Unsteady flow

An extract from the results for pulsatile flow showing the axial velocity at a number of cross-sections is shown in Figure 4. There is a very small region of recirculation on the "bed" of the artery (which is difficult to see in greyscale), but a pronounced reversal of the flow at 1 diameter downstream of the "toe". This reversal appears in the image as a concentrated dot on the inner wall of curvature.

4.3 Residence time

As reported in [12], the separation region near the toe seems to develop at two separate locations, before merging into a larger zone. This effect may explain the form of the residence time map for this part of the cycle shown in Figure 5. The recirculation region on the artery bed opposite the graft inflow is also clearly shown as a (dark) region of long residence time.

5 Modelling damage due to flow shear

The flow exerted shear stress acting on a marker particle is calculated by interpolation of the values at the surrounding nodes of the cell in which the particle is found. The criterion used to determine the stress acting on the particle is that of von Mises, specifically:

$$\sigma_e = \frac{1}{\sqrt{2}}\left\{(\sigma_{xx}-\sigma_{yy})^2+(\sigma_{yy}-\sigma_{zz})^2+(\sigma_{zz}-\sigma_{xx})^2+ 6({\sigma_{xy}}^2+{\sigma_{yz}}^2+{\sigma_{zx}}^2)\right\}^{\frac{1}{2}} \tag{5.1}$$

where σ_{xx}, σ_{xy} etc. are the components of the viscous stress tensor $[\sigma]$, which for a Newtonian fluid is given by,

$$[\sigma] = \mu[\nabla\vec{\mathbf{u}}] + \mu[\nabla\vec{\mathbf{u}}]^{\mathbf{T}}. \tag{5.2}$$

The above measure has been proposed [13] to model the response of erythrocytes and platelets, where transient exposures to high shear stress can be survived, although sustained levels of moderate shear are damaging. Although there is some data on shear stress exposure for erythrocytes, at present there is a scarcity of comprehensive data for platelets, partly as a consequence of the apparently greater sensitivity of platelets. For the present we have implemented a simple prediction of accumulated shear-induced load.

5.1 Unsteadiness and separation effects on transport at high Péclet numbers

Recently the non-linear dynamics of high Péclet number transport in unsteady flows has been studied. Chaotic mixing of trajectories is a common feature of many flows, which we think may be important when modelling particle transport and adhesion. In [14] an approximate solution of the 2D Navier-Stokes equations in the vicinity of a temporally-perturbed separation zone was studied, using a stream function defined by:

$$\psi(x,y,t) = -\frac{y^2}{2} + y^3 + \frac{x^2y^2}{2} - \frac{y^4}{6} + \beta xy^3 sin(\omega t) + O(5). \tag{5.3}$$

where β relates to the perturbation amplitude. To illustrate effects of unsteady separation on particle transport at high Péclet numbers, we examine solutions of the above at a frequency $\omega = 1$, for a thin stream of particles approaching the separation zone, Figure 6. If the flow were steady, the particles would bypass the separation, (a non-hyperbolic stagnation point) however as shown in Figure 7 the unsteadiness entrains particles into the recirculation zone for extended periods. The apparent structure of the particle paths is a consequence of the dynamics of the unstable manifold for the map of the stagnation points. Introduction of small diffusivity ($Pe = 5 \times 10^{-8}$) (using the random walk method) shows that the essential features of the non-diffusive solution are still present with small diffusivity.

Within the context of our studies, the above serves as a useful test case for procedures to investigate the spatio-temporal origin of particles adhering to the vessel walls. In Figure 8(a), the dependence of the trajectories arriving at a fixed point on the instant of release during the flow cycle is illustrated for a few trajectories. In Figure 8(b), the Wiener bundle method [14] is used to examine the effect of small diffusivity on particles arriving at a given instant at a given point. In this case, summing over a large number of such trajectories (the bundle) allows one to construct the spatial probability distribution for particles arriving at a given location.

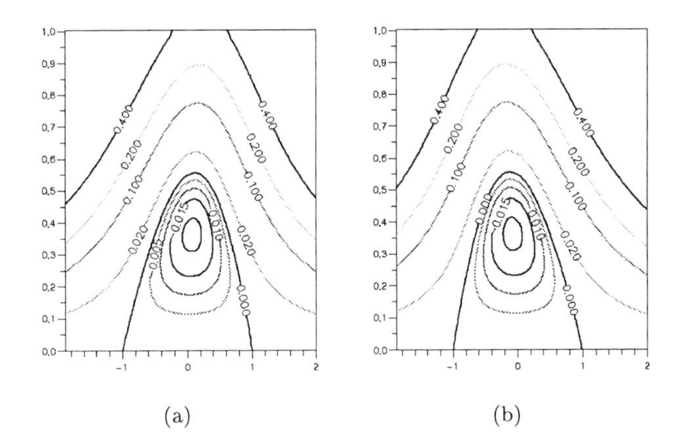

(a) (b)

Figure 6. Streamlines of weakly unsteady separation bubble as introduced by Ghosh et al. (1998) at oscillation extremes. $\beta = 0.2$

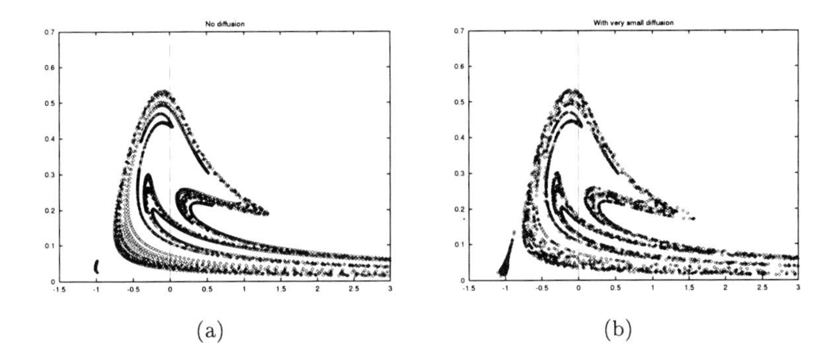

(a) (b)

Figure 7. Evolution of thin line of particles introduced initially in the interval [-2.0,-1.05] at time 40. (a) No diffusion, and (b) Weak diffusion: high Péclet number $< 10^{-7}$. Note similarity; the small diffusion effect is noticeable near separation point at (-1.,0). (Trailing portion of particle stream near separation point will eventually be swept up by flow)

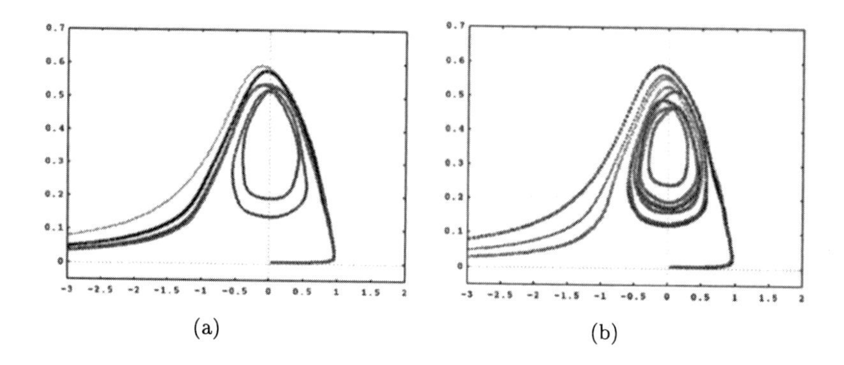

Figure 8. Trajectory of particles which arrive at (0.05,0.005). (a) Particles which arrive at the given point at different times in flow cycle, no diffusion, and (b) A few elements of a Wiener bundle of trajectories which arrive at this point at a particular time in cycle considering difusion

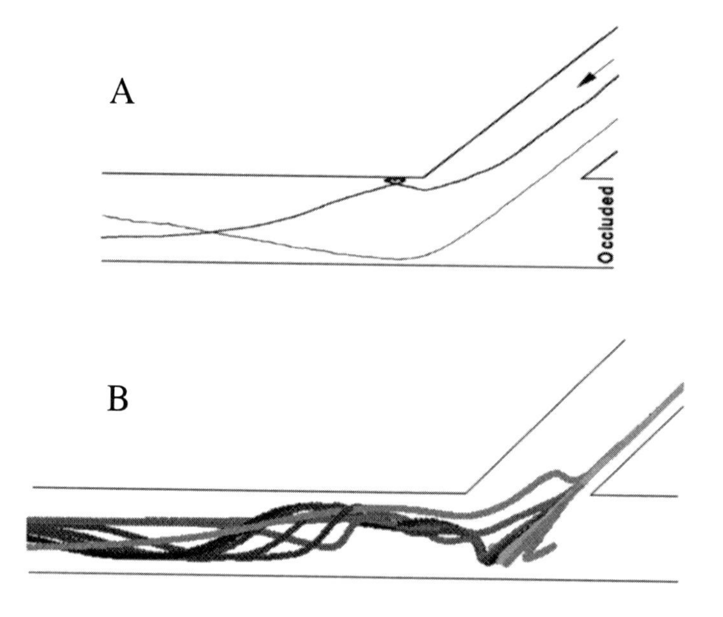

Figure 9. (A) Trajectories of two particles in steady flow through planar graft at Re 290, for Sc=100, and (B) Particles released from same point in unsteady inflow at different instants in cycle, with no particle diffusivity

Clearly this shows that the flow pulsatility as well as the topology of separation is likely to be of considerable significance for particle transport.

Figure 9(A) shows two trajectories computed for particles in a steady flow of Re 290 through the planar model graft. Here low diffusion (Sc = 100) is sufficient to allow a particle to enter the small recirculation region downstream of the toe, and eventually exit. The lower trajectory is for a particle which accumulates high shear exposure in passing close to the reattachment point.

Figure 9(B) shows trajectories of particles released from the same point in the graft inlet but at different instants in a periodic unsteady flow (as in Figure 4). Although for this case there is zero particle diffusion, unsteadiness of the complex flow produces mixing in the downstream region.

6 Numerical simulation of magnetic resonance flow measurement

An interesting application of the particle-in-cell method is to predict the signal obtained by magnetic resonance imaging (MRI) techniques used to measure flows. Many of the recent results for the geometry and flow in arteries have been obtained using MRI. Measurements made by MRI methods may however be compromised (with signal loss, distortion etc.) by certain features of the flow acting in conjunction with the particular magnetic gradient pulse sequence selected. Full numerical computation of both the velocity and magnetisation fields enables one to predict how flow patterns and pulse sequences affect the accuracy of MRI flow measurements. This should be of benefit in better relating computational modelling to in-vivo measurements, and also in the design of MR techniques.

For a good introduction to MRI, a text such as [15] may be consulted. The signal in MRI of the human body originates from the nucleus of the hydrogen atom, specifically from the magnetic spin associated with the proton which comprises the nucleus. When placed in a strong magnetic field $\vec{B}_o$ (typically of order 1 Tesla), application of a pulse of radio frequency excitation can alter the magnetisation state defined by the assembly of spins constituting the sample. Excitation energy absorbed by the spins rotates the vector $\vec{M}$ (denoting the magnetisation of a small region) out of alignment with the steady field. This causes $\vec{M}$ to precess about the primary field axis at a frequency (Larmor frequency) proportional to the instantaneous field. The spins decay back to realignment, releasing energy at this precessional frequency thereby providing a signal. By varying the field strength in space the origin of this signal may be pin-pointed from its frequency. Furthermore for spins transported by a flow to a region of different field strenth the precession frequency will vary. Thus flow velocities can be related to the differences in phase between signals from spins transported with a flow and from stationary spins.

The first stage in a numerical solution is to discretise the region of interest, and assign values for the initial spatial distribution of magnetisation $\vec{M}(x, y, z, 0)$.

The Bloch equations are a set of ordinary differential equations which govern the temporal evolution of the magnetisation ($\vec{M}$) of a stationary assembly of spins in a static field $\vec{B}_o$. For a flowing "spin" (strictly an elemental sample volume of $\vec{M}$ which moves with the flow), the temporal derivative in the Bloch equations is simply replaced by the convective derivative $\frac{\partial}{\partial t} + \vec{V} \cdot \nabla$ and one obtains

$$\begin{aligned}
\frac{\partial M_x}{\partial t} + \vec{V} \cdot \nabla M_x &= \gamma(M_y B_{gz} - M_z B_{y1}) - \frac{M_x}{T_2} \\
\frac{\partial M_y}{\partial t} + \vec{V} \cdot \nabla M_y &= \gamma(M_z B_{x1} - M_x B_{gz}) - \frac{M_y}{T_2} \\
\frac{\partial M_z}{\partial t} + \vec{V} \cdot \nabla M_z &= \gamma(M_x B_{y1} - M_y B_{x1}) - \frac{M_z - M_0}{T_1}
\end{aligned} \tag{6.1}$$

where the static and gradient magnetic fields are $(0, 0, B_o + B_{gz})$, the magnetic component of the radio frequency excitation is $(B_{1x}, B_{1y}, 0)$, (B_{1y} is usually taken as zero) and M_o is the equilibrium value of $\vec{M}$.

The evolution of $\vec{M}$ can thus be computed by tracking the spin particles along pathlines of the flow, whilst integrating the effects of the instantaneous fields experienced by the particles. As before, the velocity $\vec{V}$ is computed on a fixed Eulerian mesh, and then interpolated to the positions of the spin particles. The convective terms $(\vec{V} \cdot \nabla)\vec{M}$ are accounted for by moving the particles, so that updating the magnetisation of a spin particle reduces to numerical integration (4th order Runge-Kutta method) along the trajectories of the terms on the right hand side of (6.1) above.

With perfect homogeneity, the local field at any particle position can be computed directly from the gradient fields. For more complex spatial inhomogeneities however, the local field can be computed at the nodes, and interpolated to the particle locations assuming the grid is sufficiently fine for accuracy.

6.1 Examples

In thin slice flow measurement (or 2D phase sensitive angiography), a pulse sequence of radio frequency and magnetic field gradients is used to measure a single velocity component in the slice plane. All three velocity components can be measured in successive scans by varying the direction of flow encoding gradients. A complete 3D volume can be imaged either by stacking a series of slices, or directly, by adding a second positional phase encoding. In gradient echo flow imaging, the echo time TE should be as short as practical, particularly when imaging flows with high velocities transverse to the component being measured. The echo time is essentially the delay between marking the spins (with the field pulses), and acquiring the signal after they have travelled. Displacement of the spins by motion in the transverse plane during the delay before the echo signal is acquired produces distortion type artifacts. The form of possible measurement

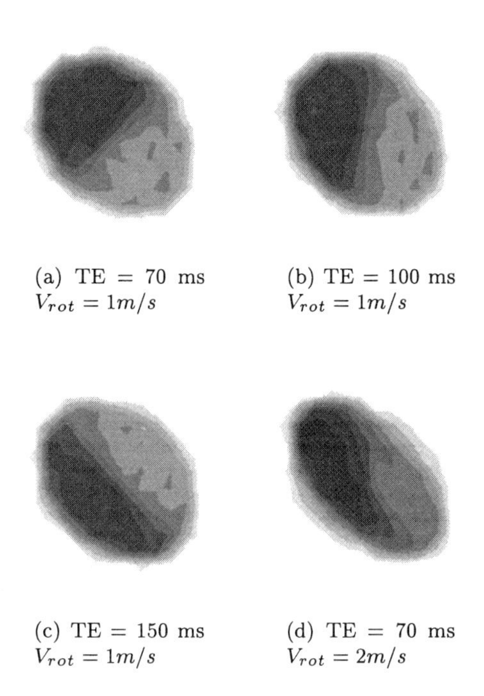

(a) TE = 70 ms $V_{rot} = 1m/s$

(b) TE = 100 ms $V_{rot} = 1m/s$

(c) TE = 150 ms $V_{rot} = 1m/s$

(d) TE = 70 ms $V_{rot} = 2m/s$

Figure 10. Effect of echo time delay on axial velocity flow image for swirling flow. The MRI image should appear as a circular section with a uniform high velocity (dark) in the upper half plane, and uniform lower velocity (light) in the lower half plane

errors is shown by applying the simulator to a model problem comprising plug flow in a circular tube, Figure 10. The axial velocity above the horizontal centreplane is uniform and twice that below the centreplane. A solid body rotational velocity is superimposed; using low velocity encoding gradients and long echo times produces large image distortion. (The low quality of the images is due to a 16×16 image matrix and simple RF excitation, however here the interest is with the gross effects).

Three effects can be seen in these images: firstly the distortion of the circular section to a skewed elliptical form, secondly the spurious rotation of the axial velocity distribution, and thirdly the smearing of the velocity jump across the interface between upper and lower velocities. Further description is given in [16].

6.2 Flow in a helix

Finally the measurement of the axial velocity component for developing laminar flow in a helix at Re 500 has been simulated, Figure 11. The CFD solution was obtained by the spectral element formulation [11] and the results for a sector at $55^o \pm 5^o$ interpolated onto a cartesian mesh of dimensions $n_x = 64$, $n_y = 64$ and

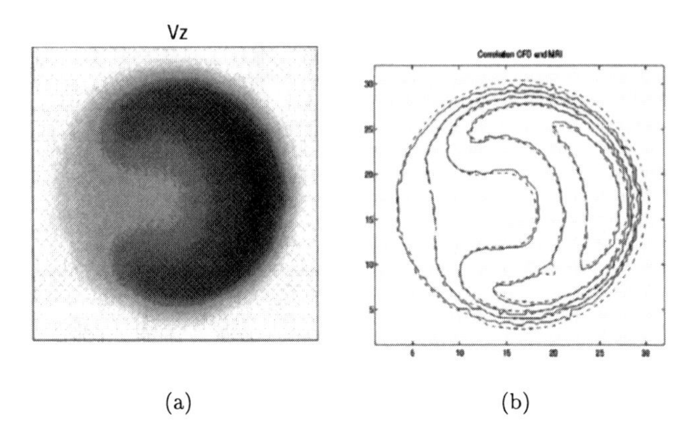

Figure 11. Simulated phase-contrast measurement of axial component of flow in a helix. (a) Simulated image, and (b) Comparison of simulated MRI measurement (dashed) and original CFD (continuous) contours of axial vleocity

$n_z = 16$. The echo time was chosen to be sufficiently short so that distortion effects were negligible. The simulated MRI measured results show very good agreement with the original computation; further results are given in [16].

Acknowledgements

The work reported here represents some of the results of a collaborative effort carried out equally by colleagues Omar Shah, Spencer Sherwin and Joaquim Peiró. We are very grateful to the EPSRC, the Bupa Foundation, the Garfield Weston Foundation and to HAT technology for their support of this work. Most of the computations were performed using the JREI and Royal Society funded Visualisation Center at the Department of Biological and Medical Systems at Imperial College.

Bibliography

1. Cussler, E.L. (1997). Diffusion mass transfer in fluid systems. Cambridge University Press.

2. Aarts, P.A.M.M., Steendijk, P., Sixma, I.J., and Heethaar, R.M. (1986). Fluid shear as a possible mechanism for platelet diffusivity in flowing blood. *J. Biomechanics*, **19**, pp. 799–805.

3. Eckstein, E.C., and Belgacem, F. (1991). Model of platelet transport in flowing blood with drift and diffusion term. *Biophysical J.*, **60**, pp. 53–69.

4. Buchanan, J.R., and Kleinstreuer, C. (1998). Simulation of particle hemodynamics in a partially occluded artery segment with implications to the initiation of microemboli and secondary stenosis. *J. Biomech. Eng.*, **120**, pp. 446–454.

5. Pozrikidis, C. (1997). Introduction to theoretical and computational fluid dynamics. Oxford University Press.

6. Chorin, A. (1973). Numerical study of slightly viscous flow. *J. Fluid Mech.*, **57**, p. 785.

7. Doorly, D.J., and Liu, C.H. (1995). Tracking the development of vortex flow structures using vortex particle-in-cell methods. *Numerical Methods for Fluid Dynamics V*, Editors: K.W. Morton and M.J. Baines, Oxford.

8. Sherwin, S.J., and Karniadakis, G.E. (1996). Tetrahedral hp finite elements: algorithms and flow simulations. *J. Comp. Phys.*, **124**, pp. 14–45.

9. Caro, C.G., Doorly, D.J., Tarnawski, M., Scott, K.T., Long, Q., and Dumoulin, C.L. (1996). *Proc. Roy. Soc., London A*, **452**, pp. 185–197.

10. Sherwin, S.J., Shah, O., Doorly, D.J., and Peiró, J. (1999). The influence of out of plane geometry on the flow within a distal end to side anastomosis. To be published *J. Biomechanical Engineering.*

11. Doorly, D.J, Peiro, J., Sherwin, S., Shah, O., Caro, C.G., Tarnawski, M., McLean, M., Dumoulin, C., and Axel, L. (1997). Helix and model graft flows; MRI measurement and CFD simulations. *ASME Paper FEDSM97-3423.*

12. Shah, O., Sherwin, S.J., Doorly, D.J., and Peiro, J. (1999). *Imperial College Aeronautics Report: Computational modelling of graft flows.*

13. Bludszuweit, C. (1994). Ph.D. Thesis, University of Strathclyde, Glasgow.

14. Ghosh, S., Leonard, A., and Wiggins, S. (1998). Diffusion of a passive scalar from a no-slip boundary into a two-dimensional chaotic advection field. *J. Fluid Mech.* **372**, pp. 119–163.

15. Gadian, D.G. (1995). NMR and its applications to living systems. Oxford University Press.

16. Doorly, D.J., and Ljungdahl, M. (1997). Computational simulation of magnetic resonance imaging. *ASME Paper FEDSM97-3422.*

The Application of Computational Fluid Dynamics and Solid Mechanics to Haemodynamics in Arterial Organs and to Related Problems

S.Z. Zhao*, M.W. Collins, Q. Long* and X.Y. Xu***

**Department of Chemical Engineering and Chemical Technology, Imperial College, London, and **School of Engineering System and Design, South Bank University, London*

(Based on Invited Lecture given by MWC)

Abstract

Cardiovascular haemodynamics forms an important current class of medical research. Over the last ten years, as the current generation of Computational Fluid Dynamics (CFD) codes have been developed, so they have been applied to arterial problems and have been demonstrated to be a valuable and reliable tool in this area. This paper will discuss a range of topics in our efforts to correlate haemodynamics to risk factors underlying the genesis and progression of cardiovascular disease. In the course of discussion, the state of art of CFD is described and its application in blood flows demonstrated. Furthermore, our approach has been extended to incorporate solid mechanics into haemodynamic studies so as to address the problems in a more realistic and comprehensive manner. To obtain predictions which can reflect the real situations in the human body, image and data processing software has been developed to take raw clinical data from Magnetic Resonance Imaging and interface it with the CFD code. As the vessels become smaller in size, so wall effects increase in complexity, and we give relevant data from a major study of nanoscale fluid problems in human biology.

1 Introduction

Cardiovascular diseases such as atherosclerosis are known to be influenced by blood flows in terms of their genesis and progression [1, 2]. In order to investigate what is a complex mechanism a broad knowledge of detailed arterial flow phenomena is necessary. Modern techniques such as ultrasound and Magnetic Resonance Imaging enable measurements of pulsatile blood velocities to be undertaken. However, accurate assessment of wall shear stress, which is of considerable physiological interest, is quite difficult. An alternative approach to measurement of detailed arterial flow characteristics employs computer simulations of vessel models which makes it possible to obtain temporally and spatially varying shear stress. Unsteady flow in a complex, possibly time varying, geometry is very complicated. The high level of computer technology and recent developments in CFD now means that 3-dimensional transient flows in irregular geometries can be simulated as a matter of course by standard commercial codes. Also features such as non-Newtonian effects and time-dependent boundaries can be accommodated. Immediate benefits have been in the field of atherosclerosis, where the haemodynamics has been commonly agreed by medical specialists to be significant, and in the design of arterial heart valves. This paper concerns the former area.

While great progress has been made in a number of studies of blood flow, it soon become apparent that the general problem was one of fluid-wall interaction, and necessitated consideration of the solid mechanics of the wall [3, 4, 5]. The stress-strain rate equations for fluids and stress-strain equations for solids are similar, and considerable effort was spent in our Research group in inserting simple wall models into two (finite volume and finite element) CFD codes. This was not without success, both attempts resulting in simulations (among others) of the Womersley problem for pulsatile flow in a compliant tube. However, wall compliance and muscle effects comprise the highest level of solid mechanics (SM) models available today in the commercial codes, and our latest work is in a successful iterative coupling of SM and CFD codes.

To make numerical simulations of clinical relevance it is necessary to provide data for both geometry and flow boundary conditions. In some funded current work, special image- and data-processing software has been developed to convert Magnetic Resonance Angiography (MRA) images into the format that the CFD code can accept by operating as an automatic interface between them both [6].

Surgery and medical research methods have progressed to increasingly smaller scales, and a funded substantial review exercise has been undertaken of the methods available to model fluid transport problems in the nanoscale range. Again, we hope to commence numerical simulation studies for a significant medical problem.

In this paper, examples will be given of the predictions and comparisons achieved.

2 Developments in CFD

The mathematical description of local blood flow uses the time-dependent, three-dimensional, incompressible Navier-Stokes equations for Newtonian and non-Newtonian inelastic fluids. In general, currently available techniques in CFD to solve the above equations fall into four categories. They are the finite difference, finite volume, finite element and boundary element approaches. The finite difference or the finite volume method was historically the first to be developed and has been available for many years. It has been established that this approach gives stable results with relatively less computation time than is required by other methods. Although the finite element method allows greater flexibility to build the computational domain, much more time and memory space are necessary for computation. The boundary element method is mainly used for a class of flow problems (potential flow) which is not relevant to blood flow.

It is recognized that the fluid-wall interaction type problem can not be properly dealt with by a finite difference or finite volume method. In most efforts so far reported to solve the coupled equations for the wall motion and the fluid motion, a finite element method has been utilized. This is partly because the solid mechanics problems have been mostly solved by finite element method.

It is well known that CFD is a powerful tool for investigating the complex nature of blood flows. Computational modelling of blood flow has the potential to provide complete haemodynamic data for flow related phenomena and any derived parameters of mechanics. This is particularly important in studies of the role of blood flow in atherogenesis because it is a combination of these parameters which seems to play an influential role in the process. Although there are many difficult problems in applying CFD to blood flows, (for example, its unsteadiness, and the non-Newtonian viscosity of blood) it now has become a matter of computing power to overcome them [7, 8]. In fact, there are a variety of CFD codes available to cover these issues.

In our first numerical simulations we used the ASTEC code [9] with its combination of unstructured gridding and finite-volume-type algorithms. ASTEC is now an option within CFDS-CFX4. When the *multi-block* version of the (then) FLOW3D code was released, we found it to be particularly effective for bifurcation studies [10]. However, we have also used the finite-element solver FEAT [11], developed in the old Central Electricity Generating Board by Dr. Tony Hutton and others.

The equation solver used in our group, then, is the well validated CFD code–now termed CFDS-CFX4–a finite volume based code using a structured, patched multi-block, nonorthogonal, curvilinear coordinate grid with a collocated variable arrangement [10]. The basic solution algorithm is the SIMPLEC pressure correction scheme. CFDS-CFX4 has models for multi-phase flow, particle transport, gaseous combustion, compressible and incompressible flows, porous media flow, and so on.

Numerical simulations of blood flow in human organs applying various releases of several CFD codes have been undertaken in our group for the last ten years with a number of projects going on in parallel. It covers the fields of arteries, bifurcations, grafts, and skeletal muscle ventricles [12, 13, 14, 15]. We initially studied the haemodynamics of bifurcations, and after a comprehensive range of code validations involving in vitro (laboratory) data for 2- and 3-dimensional T-junctions, we made simulations for in vivo clinical cases of femoral bifurcations and in vitro end-to-side anastomoses [16, 17]. As the moving boundary capability of CFD became available, we extended our work to study the blood flow in skeletal muscle ventricles (SMV) which is a wall-driven unsteady flow with a large volume change and time-dependent gridding [18]. In general velocity predictions compared well with measurements.

All the above studies, however, have either assumed a rigid wall or at most prescribed wall movement from experimental data. To obtain a much clearer understanding of physiological flows and wall behaviour, it is recognized that the blood-wall interaction cannot be neglected in such cases as the cardiovascular system. In the subsequent projects, our emphases were placed on the fluid-wall coupled situation which will be discussed below.

3 Application to bifurcations

The arterial system carries out its function of distributing blood throughout the body by means of a remarkable network characterized by vessel branching and bifurcation, and fluid dynamics patterns at these sites of flow division can be extremely complex by comparison with flow in nonbranching vessels. In certain larger and medium-sized arteries, such as the carotid, coronary and aorto-iliac arteries, these sites of branching and bifurcation are associated with the development of atherosclerotic plaques, and haemodynamic factors such as wall shear stress and particle residence time have been implicated as participants in atherogenesis [2]. The complex flow fields which exist in the region of arterial bifurcations are characterized by strong spatial and temporal variations in wall shear stress, creating environments considerably different from those found in simple systems. In addition to the natural bifurcations occurring in the arterial system, "artificial" bifurcations may be created during vascular graft surgical procedures that bypass arterial obstructions. In such cases the specific geometry of the anastomosis of the graft to the host artery is determined by the relative diameters of the graft and artery and by local anatomy.

3.1 General features of flow in bifurcations

From the theoretical viewpoint, when applying the governing equations of motion, the boundary conditions associated with bifurcating flows are the primary determinants of the flow behaviour. Hence, the geometry of the bifurcation, the inflow conditions, and division of flow ratio all play strong roles in dictating details of the local flow field.

The fact that flow changes its direction as it moves through the bifurcation means that it acquires a secondary flow in the branches. However, flows in arterial bifurcations, in addition to having secondary flow patterns due to curvature, are also affected by the rapid increase in cross-sectional area which is typically encountered, producing locally adverse pressure gradients that influence near-wall velocity profiles, particularly along the outer walls. This region is also one in which there is a relatively long residence time of particles. We have studied a number of different bifurcation geometries, but for reasons of space give results for a current project only.

3.2 Flow in a human carotid bifurcation model

The human carotid bifurcation is unusual among the systemic arteries because of the presence of the carotid sinus in many subjects and because of the low distal resistance of the vascular bed supplied by the internal carotid branch, resulting in a continuously forward flow even during diastole and in a time-varying flow division during the cardiac cycle. The carotid bifurcation is important clinically because of its predilection to develop atherosclerotic plaques which can lead to transient ischemic attacks and strokes due to plaque ulceration and embolic events.

As currently we are involved in a project about the acute effects of oxygen and carbon dioxide on carotid haemodynamics, six human carotid bifurcation models were reconstructed for the study from ultrasound images by assuming planar bifurcation. Due to space limitations, the results of only one of the six models are discussed here in order to demonstrate the basic characteristics of flow patterns found in human carotid bifurcations. All CFD results presented here were taken from the third cycle of the calculation. Internal validations have been performed by grid and temporal density tests to guarantee grid- and time-step independence of results.

Fully developed velocity profiles corresponding to the time-dependent mass flow rate at the entrance of the common carotid and the flow division ratio between the internal and external vessels were prescribed as boundary conditions. Figure 1 shows the flow division ratio between the internal and external carotid from pulsed ultrasound Doppler measurements. It demonstrates a widely varying value of the ratio over the cardiac cycle, in contrast to the majority of previous studies where a division ratio of 70:30 has usually been employed.

Figure 2 presents a schematic diagram of the model geometry and of the major flow features such as the flow separation zone, with very strong skewing velocity toward the inner wall and secondary flow in the branches. Under pulsatile flow conditions the separation zone is transient both in size and in location. However it does persist over a substantial portion of the cardiac cycle, in this case occupying 40% of the cycle. Shown in Figure 3 are the wall shear stress variations at selected locations on the model. As expected from the velocity profile behaviour, the wall shear stress distribution in the bifurcation model shows the outer wall of the sinus to be a region of low mean shear and of rather small oscillations in magnitude and direction, and the inner wall to be a region of relatively high wall shear which oscillates only in magnitude. For clarity we have concentrated on the symmetric plane in spite of the 3D configuration.

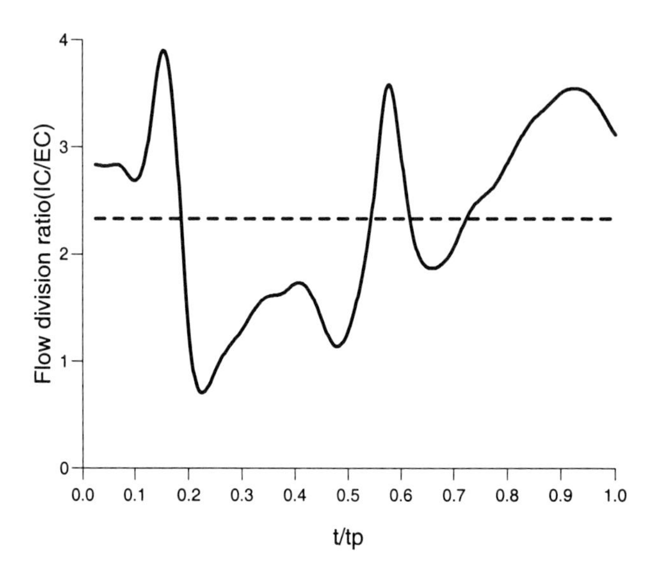

Figure 1. Time-varying flow division ratio of internal and external carotid arteries (dashed line corresponding to a constant ratio of 70:30)

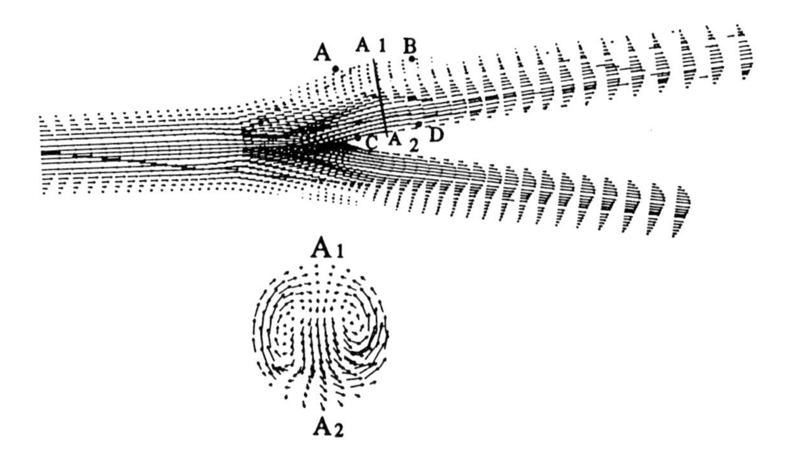

Figure 2. A typical velocity vector and secondary flow pattern in a human carotid bifurcation

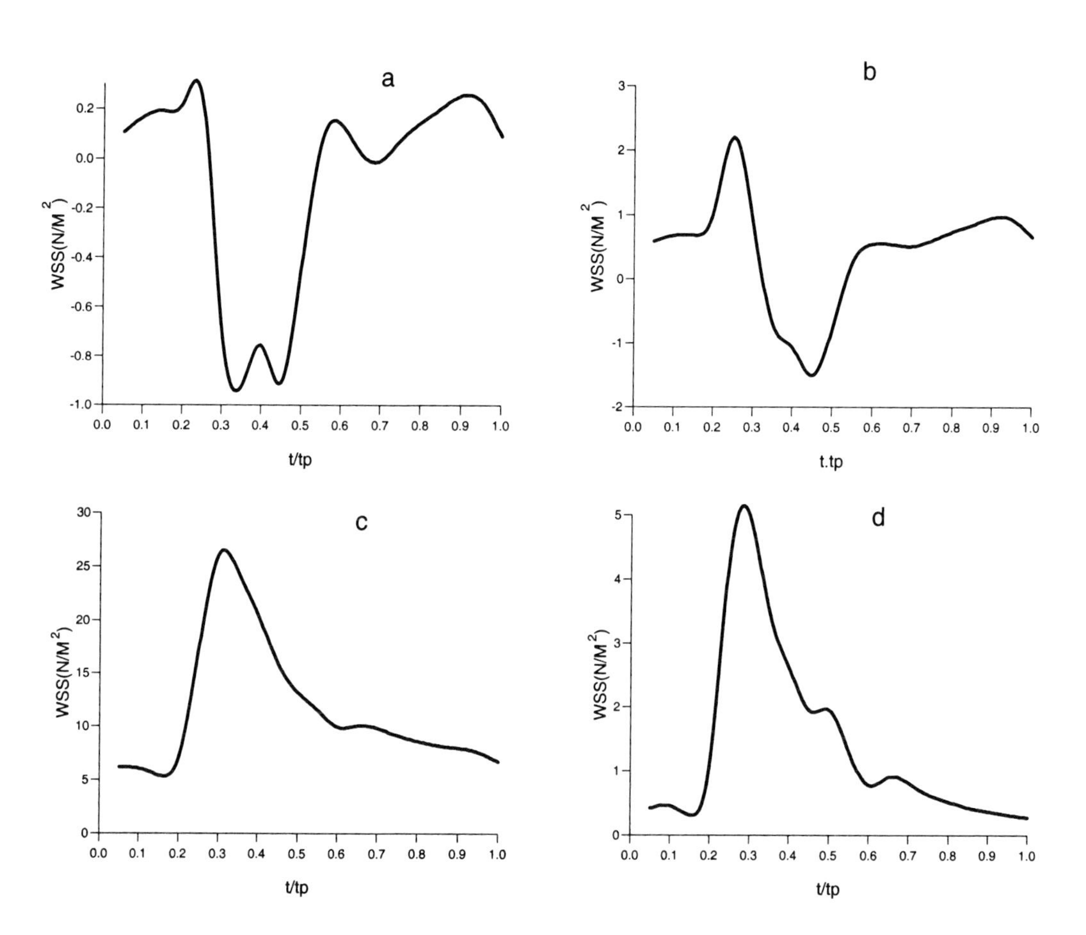

Figure 3. Wall shear stress variations during a cardiac cycle at four selected sites of a carotid bifurcation (as indicated in Figure 2)

4 Incorporation of solid mechanics models

4.1 Initial developments: Incorporation of wall effects in CFD codes

Our initial efforts were based on the similarity of the equations of motion for both a fluid and a deformable wall. The Navier-Stokes equations for an incompressible fluid are:

$$\mu_f g^{jk} u_{i;jk} - \frac{\partial p}{\partial x^i} = \rho_f (a_i - b_i) \tag{4.1}$$

where μ_f is the viscosity, g^{jk} a metric coefficient, u_i the velocity vector, p the pressure, ρ_f the density of the medium, a_i the acceleration vector, and b_i the body force vector per unit mass. Similarly the equations of motion for a linearly elastic, homogeneous, incompressible, isotropic material are:

$$\mu g^{jk} \xi_{i;jk} - \frac{\partial p}{\partial x^i} = \rho (a_i - b_i) \tag{4.2}$$

where μ =2E/3, E the Young's modulus, ξ_i the displacement vector, and ρ is the density of the medium.

Considerable efforts were made by Drs. Frank Henry and Yun Xu to insert simple wall models into the algorithms for the finite volume CFDS-CFX and the finite element FEAT codes respectively. A number of code validation exercises were undertaken, notably the Womersley solution for pulsatile flow in an elastic tube. These are reported in Section 7.

4.2 The coupled wall-fluid approach and its application to pulsatile flow in a real human carotid bifurcation

Although it is now recognized that specific patterns of local blood flow predispose the development of atheroma, the mechanisms underlying the important determinant of cardiovascular risk are incompletely understood. It is believed that more than one mechanism may be involved in atherogenesis, hence other possible factors are currently being examined. One area that is gaining attention is the role of mechanical stress on the vessel itself in atherogenesis. At the same time, however, theories persist that haemodynamic related factors are the main contributors to lesion formation. Much solid mechanics oriented work has been done via the stress concentration studies of the arterial wall, loading the wall with a normal incremental pressure of, (usually) 40 mmHg. This corresponds to the pressure difference between peak systolic and diastolic pressure phases. Historically, the development of CFD and SM codes have been developed separately, and in parallel. Perhaps the concept of the duel roles of fluid dynamics and solid mechanics may lead to further insight of the mechanism underlying the formation of lesions by treating the system as a coupled one.

Mathematical analysis of arterial flow including wall motion is only possible for the simplest of cases due to the requirement that the equations of motion for the flow and the wall have to be solved simultaneously. Most authors have used either self-developed packages or improved commercial packages. The various assumptions and approximations made in these studies preclude them from being extended to complex geometries such as bifurcations [19, 20].

Alternative approaches, then, are needed to treat the coupled system of equations governing the flow and wall motion in arbitrary geometries. In our group, a novel numerical method combining two commercial codes for coupled fluid-solid problems has been developed. The CFD and SM codes used are CFX4 and ABAQUS respectively. In a coupled approach the wall equations and the flow equations subject to the moving boundaries are solved. The total inner wall pressure resulting from the Navier-Stokes calculation constituted the load. From previous rigid wall calculations it is known the viscous stresses are very small compared to the pressure and can be ignored. The transient wall calculation yields a deformation vector in each surface node. With respect to these vectors the boundary of the flow domain and computational grid are updated. A full description of our iterative algorithm can be found in Zhao et al. [21, 22]. The main characteristic of our method is to ensure proper interaction between the two codes during each pulse cycle in the time-dependent calculations. The method is able to predict the full, time-dependent displacement and stress fields within the wall, as well as the details of the flow field.

Numerical predictions have been carried out for physiological flow in a real compliant human carotid bifurcation model using our developed coupling algorithm. The geometry and computational grid of the carotid model were generated from non-invasively obtained in vivo MRI data of a healthy human subject (Figure 4). Details of the grid generation can be found in Long et al. [6]. It can be seen that there is no symmetric plane at all in this realistic bifurcation. Flow waveforms obtained by pulsed ultrasound Doppler were applied as the time-dependent boundary conditions. The flow pulse waveforms used at the internal and external carotid outlets as well as the pressure pulse waveform at the common carotid inlet are shown in Figure 5.

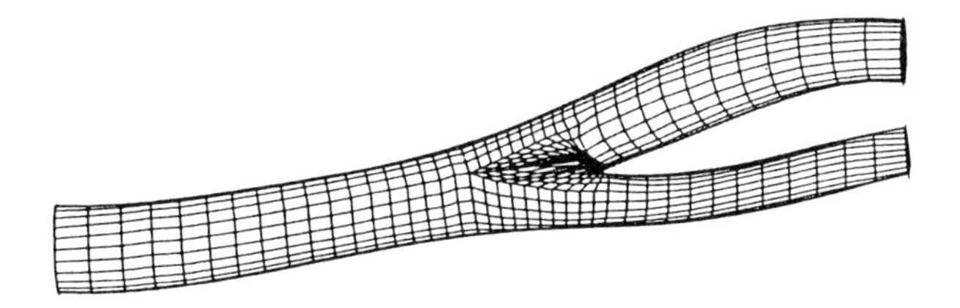

Figure 4. The geometry and computational grid for the coupled model

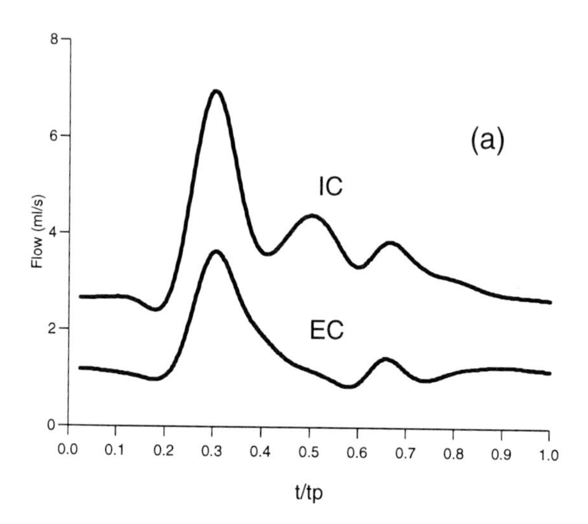

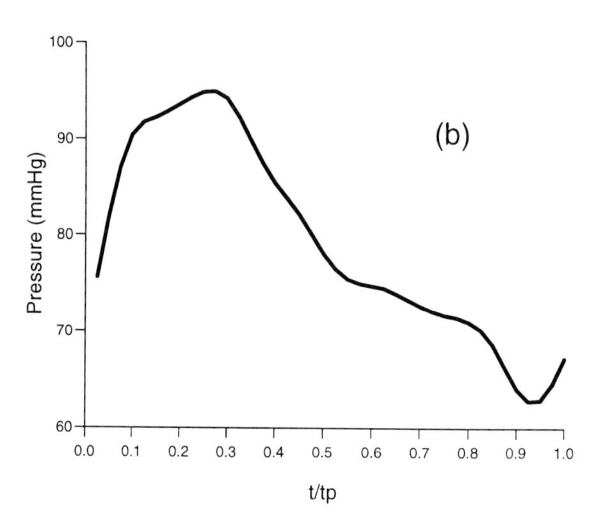

Figure 5. Flow waveforms of internal and external carotid arteries and pressure waveform of the common carotid used as boundary conditions for the coupled model

4.3 Results and discussion

The coupled calculations provided a complete set of haemodynamic data including velocity, wall shear stress distributions in the flow domain and mechanical data including displacement and intramural stress distributions on the arterial wall during the whole cardiac cycle. It is rather difficult to present the CFD results due to the non-planarity of the bifurcation and the pulsatility of the flow. Here only some typical results are shown.

Figure 6 shows the velocity contour and secondary flow at the flow deceleration phase in the bulb region in particular. Consistent with the previous finding, a skewed velocity towards the inner wall and very strong secondary flow can be found in the bulb area. However, in contrast to the symmetric bifurcation case in Section 3, the secondary flow shows a rather asymmetric nature.

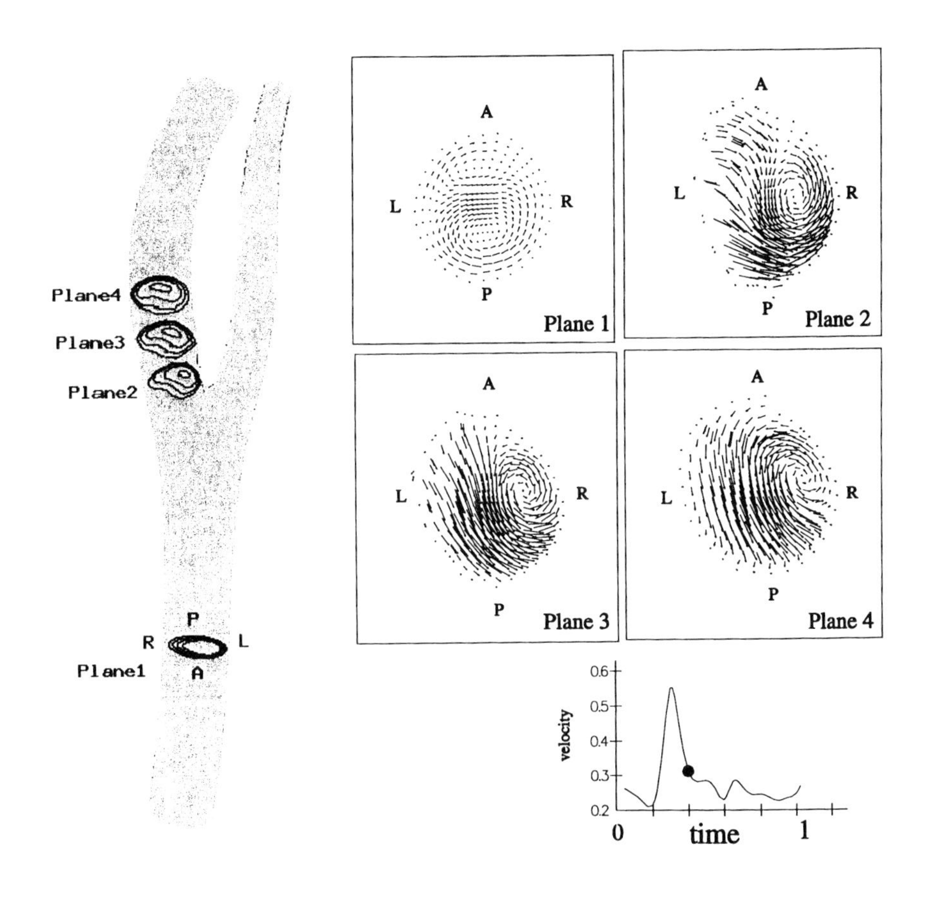

Figure 6. Velocity contour and secondary flow patterns at selected planes for the coupled model

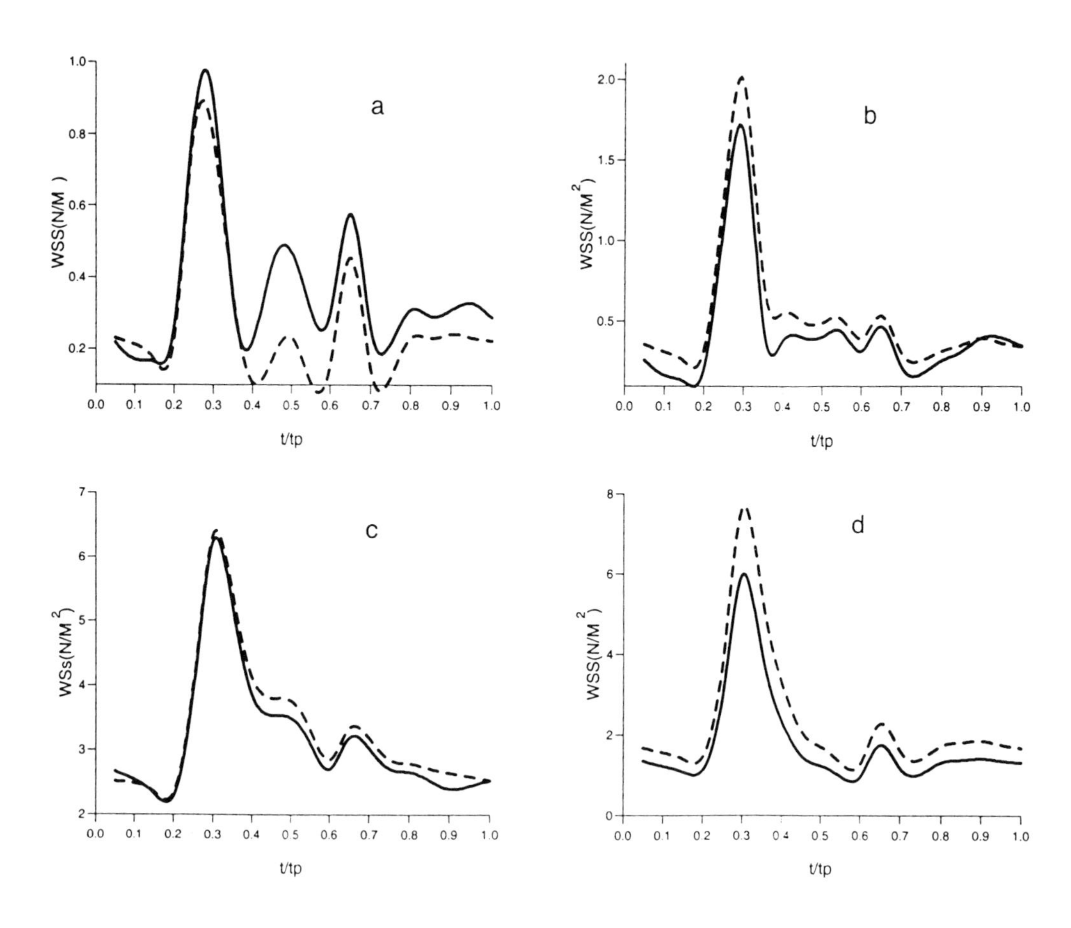

Figure 7. Comparisons of wall shear stress at four selected locations ((a) and (b) along the outer wall of the bulb, and (c) and (d) along the inner wall of the bulb) of the coupled model and its corresponding rigid model. Solid line–coupled model; dashed line–rigid model

The quantitative influence of the wall distensibility on wall shear stress is demonstrated in Figure 7. The vessel wall mechanics are analysed by means of displacements and principal stresses. Figure 8 shows the displacement distribution at the pulse phase angle of maximum pressure and the displacement variations during one cardiac cycle at the maximum displacement point which is located at the junction of the two daughter vessels. From Figure 8(b) it can be seen that the curve follows that of the pressure waveform (Figure 5(b)). This demonstrates that the pressure waveform plays the main part in driving the wall movement while the pressure gradient resulting from the flow only has a secondary influence. The maximum displacement is observed at the intersection between the two branches. Figure 9 is a plot of the maximum principal stress distribution at the inner surface of the wall at maximum pressure load. The right hand side diagram (b) is obtained by turning its counterpart by 180°, so that together they give a complete view of the stress distribution. The stress concentration factor is defined as the ratio of the local stress to the uniform stress which in the common carotid is 4.19 in the bulb region. It can be noticed that

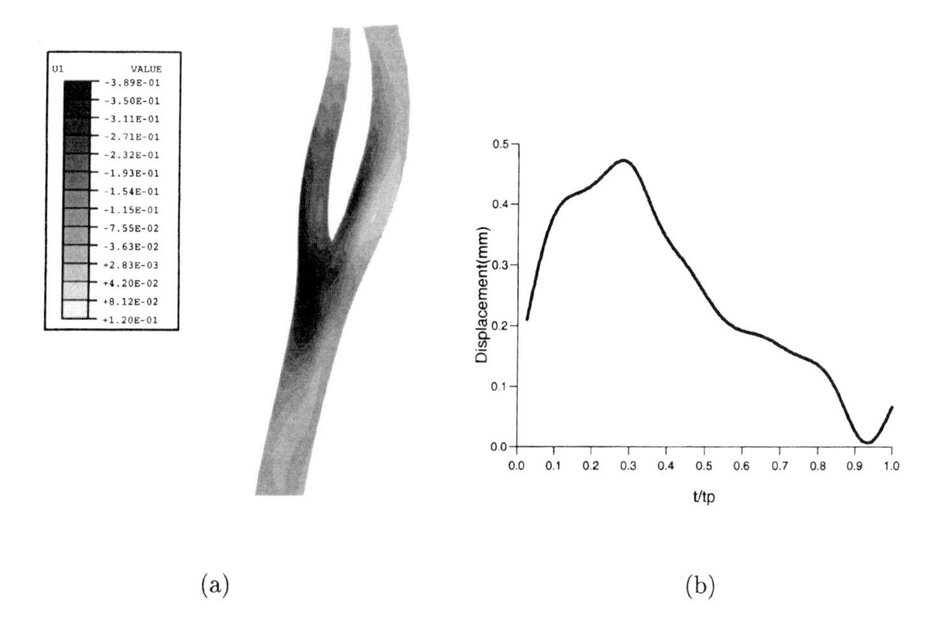

Figure 8. (a) Displacement distribution (in mm) at the maximum pressure pulse phase of the coupled model, and (b) the displacement variations during one cardiac cycle at the maximum displacement point

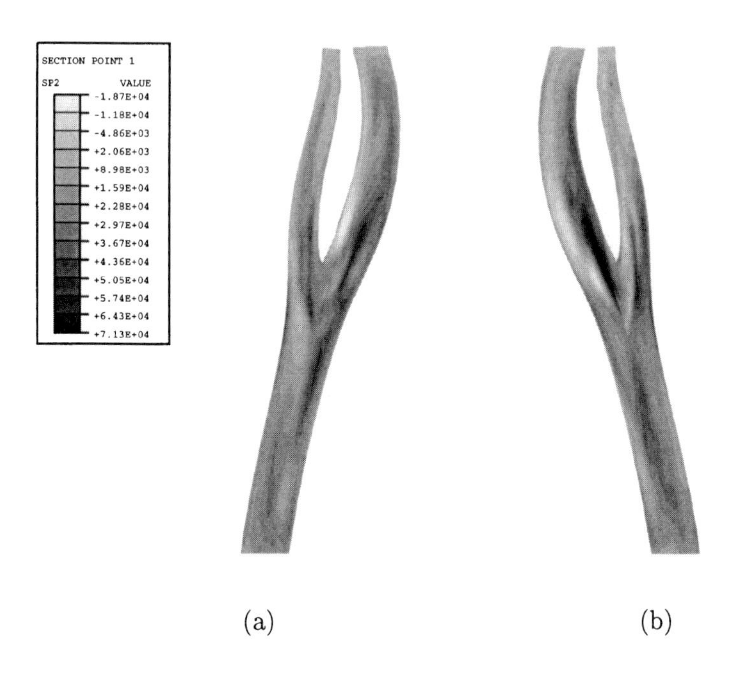

Figure 9. Maximum principal stress distribution (in Pa) on the inner surface of the coupled model at the maximum pressure pulse phase

the maximum principal stress occurred at the side wall of the bulb, while a very low stress distribution was found at both the outer and inner walls of the bulb. In addition to the locally high stresses, steep stress gradients can be observed.

The study shows that the distensibility of the vessel wall in the physiological range affects the flow field quantitatively. This also has been confirmed in the studies by Anayiotos [23], and Reuderink [24]. Generally the wall shear stress magnitude is reduced in the compliant model. Further studies are necessary to obtain a sufficient understanding of the influence of the wall distensibility in atherogenesis and in fact whether this influence is important. Wall distensibility causes wall motion and wave phenomena; the wave is reflected at discontinuities. In our model local reflection is not included. However, it should be appreciated that the most essential discontinuities in the arterial system are vessel bifurcations, and that according to Pedley [25] local reflection at bifurcations is small.

In the current model incrementally linearly elastic behaviour of the vessel wall has been assumed. The validity of this assumption in local analysis has been proven in various calculations by application to the case of a pressure loaded cylinder having different mechanical properties: linearly elastic, hyperelastic and viscoelastic behaviour. Comparison of the radial displacements indicated minor influence of hyperelasticity and viscoelasticity on local wall deformation. The modelling of the vessel wall properties is essential in global (system) investigations to give pulse propagations. In local calculations the preceding simplification is acceptable.

Although both geometry and flow are of complex nature in the human cardiac system, the coupled approach discussed here should in principle be applicable to flow problems such as in the left ventricle.

5 Development of automatic data generation software

Though the potential of CFD is very high, its applications are still limited mostly to arterial and cardiac flow models of rather simple geometries. As is frequently pointed out, the most significant factor which affects the flow structure and hence the disease is the 3D configuration of the flow field [26, 27]. Therefore the fidelity of the computational model is crucial for correlating fluid mechanics to physiological and pathophysiological phenomena such as atherogenesis. Recent developments in MR techniques have considerably improved temporal and spacial resolution, making non-invasive imaging of complex human vessel structures both possible and practicable in clinical routines. In addition, MRA offers the possibility of quantification and visualisation of blood flow in the human body non-invasively which will be especially useful in the generation of CFD boundary conditions. Crucially, it is the ability to generate images for both structure and velocity fields, which gives sufficient input data to run a CFD simulation.

Software has been developed to process MRA images, generate anatomically realistic models and construct velocity boundary conditions for CFD treatment of arterial bifurcations. In the research, serial cross section images of arterial vessels were acquired by conventional 2D Time-of-Flight (TOF) MR scan. The software developed is able to process these images and generate a numerical mesh in the bifurcation model automatically. In doing so, firstly the cross-sectional images were segmented by a program based on the *snake* model [28]. Secondly, the bifurcation model was reconstructed to 3D by serial smoothing and interpolation steps to ensure a quantitatively reliable model which could demonstrate most of the features of the arterial bifurcation. Figure 10 shows the effect of the 3D smoothing algorithms. The surface information for the model (surface point locations) was presented in well organized (structured) surface splines as shown in Figure 11. Finally, a structured numerical mesh was generated in the bifurcation; this mesh serves as an interface with the CFD code [6]. A typical numerical mesh of a human right carotid bifurcation has already been shown in Figure 4.

For the CFD boundary conditions, 3D time-dependent velocity profiles were measured by phase-contrast MR scan in inlet and outlet boundary planes of the arterial bifurcation. Software was developed to treat those velocity images semi-automatically. The velocity boundary conditions were generated by registering

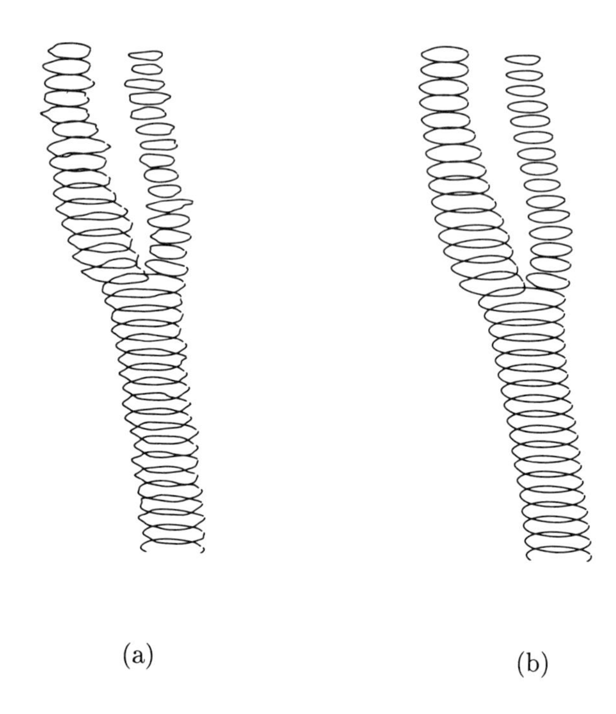

(a) (b)

Figure 10. The cross-section contours before and after smoothing

and interpolating these images in the spacial (to each boundary grid point) and temporal (to each time step of the CFD calculation) directions [29].

Compared with MRA, Ultrasound Angiography (USA) can also supply arterial vessel images and make measurements on velocity distribution in chosen planes. USA always has higher spacial and temporal resolution than MRA imaging. It can even offer on-line real time imaging. This makes vessel wall movement tracking possible which is very useful in generating a *time-dependent* vessel or ventricle geometry. Although signal loss at the vessel wall may cause problems in image processing, the newly-developed power image and different scan angle or 3D special compound technique can compensate to some extent [30]. The main drawback of USA is the difficulty of 3D registration. To solve this problem, additional devices are needed, such as a position transducer. Considering its low price of diagnosis, its wide availability and its high spacial and temporal resolution, research on using USA for 3D numerical simulations of blood flow could eventually lead to new tools for obtaining quantitative haemodynamics data in a clinical environment.

The most important feature of our work is that an automatic interface has been established between MRA and the CFD code by taking direct clinical data. This work is being extended to corresponding USA information. For the other

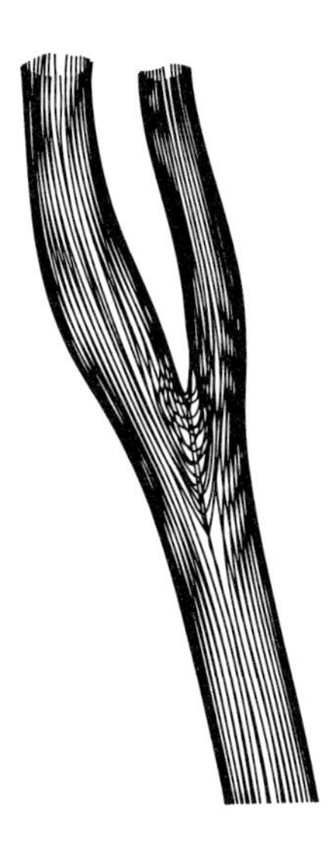

Figure 11. Surface splines rendering for the generated bifurcation

(post-processing) end of the CFD code we hope to start soon a clinical collaboration of a three-way nature, involving an engineering consultancy company with special Virtual-Reality type software. The objective is to maximise the clinical value of the large amounts of data generated.

6 Nanoscale transport processes

Physiological flows present an almost overwhelming variety of problems involving practically all scales and all common states of matter. Most problems which have received consideration relate to large-scale, macroscopic phenomena like respiratory flows and blood circulation. However, studies concerning microscale and nanoscale aspects of human physiology are becoming more and more common. In a cubic centimetre of, say, pulmonary tissue, gas, liquid and solid phases are simultaneously present, in addition to colloids, mucus, ions, viscoelastic membranes, and all kinds of macromolecules. At the nanoscales the molecular machinery of living tissue is incessantly involved in performing transport processes which range from convection to diffusion, vesicular transport, and chemical interactions. Although our knowledge of microscopic anatomy of living tissue has progressed enormously since the introduction of electron micrography, progress concerning the microscopic physiology has been much slower. The simultaneous application of advanced modelling and computational methods and of high-resolution, non-invasive diagnostic techniques, will certainly help in advancing our knowledge of the dynamics of nanoscale physiology [31]. Computational microhydrodynamics can play a central role in this effort. Figure 12 shows the multiple scales in mass transport from large vessels to the individual cells.

Endothelial cells form a tight lining on the inner wall of continuous capillaries and of most large and medium-sized blood vessels. Most exchanges between the blood circulation and the tissues occur through the wall lining of capillaries. The importance of a correct understanding of the transport process at this level is therefore self-evident. It has been recognized for several years that various types of specialized cells respond to the flow environment in which they reside, and, in particular, to fluid-induced wall shear stress. For example, in vitro studies on cultured endothelial cells from arterial walls have shown that the shear stress induces complex changes in their shape and functions [32]. All of these phenomena are of great importance for our understanding of atherogenesis and other pathologies. However, the mechanisms by which the cell senses its fluid environment and transduces mechanical signals into chemical signals are not fully understood. The most serious problem presented by modelling nanoscale physiological flows is that different effects (hydrodynamic, chemical, electrical) are simultaneously present, and often cannot be separated and treated independently. Therefore, nanoscale fluid dynamics is a highly interdisciplinary field which requires competence from medical, biological, chemical and mechanical areas to be put together.

We have been involved in a comprehensive review study of nanoscale fluid dynamics in physiological flows. Our immediate purpose has been to select a number of flow and transport problems, occurring in human physiology, in which nanoscale aspects play a crucial role, and to conduct a state-of-the-art review of existing models and computational tools applicable to these. In the future,

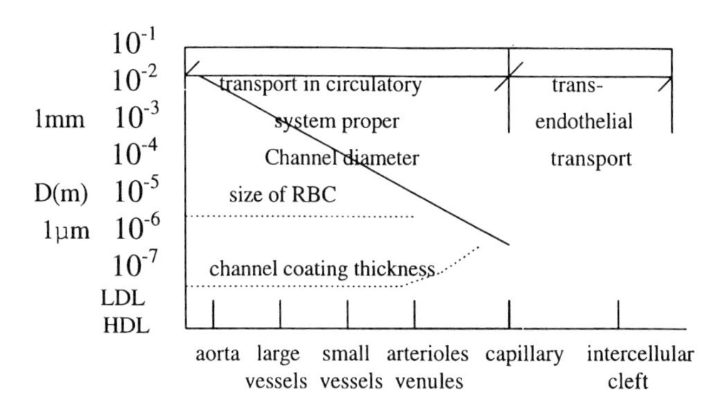

Figure 12. Comparison of dimensions of blood vessels, wall effects and blood constituents

we expect to develop specific research programs in this area, concentrating on those problems on which the keenest interest will be registered and in which computational methods will appear to be most useful. In the context of this paper, current evidence is that conventional fluid dynamics is applicable down to some of the smaller vessels where microhydrodynamics becomes the modelling approach. Even here, however, nanoscale range effects probably form an interface which should be considered.

7 Resume of research programs

Over about the last decade our research group has studied a substantial number of arterial problems. In the process various critical reviews have been made, and a variety of code validation exercises undertaken. Because of space limitations these are summarised in the form of Tables 1 and 2, as shown, together with relevant publications. For the sake of completeness, the fluid-solid interaction validation cases c–f (Table 3) are given in Figures 13–16 respectively.

Table 1. Arterial problems studied

Problem	Research Worker	Reference
ARTERIES a)Carotid artery (EDRF-related intervention)	X.Y.Xu	[14]
BIFURCATIONS b)Femoral canine(ultrasound in vivo) c)Human carotid(MRI in vivo) d)Human abdominal	 X.Y.Xu Q.Long Q.Long	[13, 14] [15, 16] [this paper] [6]
GRAFTS e)Proximal and distal(in vivo cases)	F.S.Henry	[17]
STENOSES f)Preliminary study	F.S. Henry	[33]
SKELETAL MUSCLE VENTRICLE g)Comprehensive modelling(in vivo cases) h)Comparison with model natural L.V.	 F.Iudicello F.S.Henry M.W.Collins	 [18] [34] [35]
STENTS i)Case studies-simulated surgical alternatives	X.Y.Xu	[36]

Table 2. Published reviews in numerical simulation area

Subject	Research worker	Reference
Bifurcations	X.Y.Xu	[12]
Fluid-Solid Interactions	S.Z.Zhao	[22, 37]
Nanoscale Fluid Dynamics	M.Ciofalo	[31, 38]
Magnetic Resonance Angiography /CFD Combination	Q.Long	[39]

Table 3. Significant code validation exercises

	Subject	Research Worker	Reference
Bifurcations	a)In-vitro T junction and code comparison(steady flow)	X.Y.Xu	[12]
	b)In-vitro T junction (pulsatile flow)	X.Y.Xu	[40]
Fluid/Solid	c)Expanding/contracting wall (decoupled)	F.Iudicello	[41]
Interactions	d)Long thick-walled cylinder (decoupled)	F.S.Henry X.Y.Xu	
	e)Finite-length thick-walled cylinder(decoupled)	F.S.Henry X.Y.Xu	
	f)Womersley problem (coupled)	F.S.Henry	[42]
	(coupled)	X.Y.Xu	[14]
	(coupled)	S.Z.Zhao	[21]
Skeletal Muscle Ventricle	g) Vortex ring travel in sigmoidal filling (in vitro comparison)	F.Iudicello	[34]
MRI/CFD	h)Reproducibility of individual patient's bifurcation geometry	Q.Long	[43]

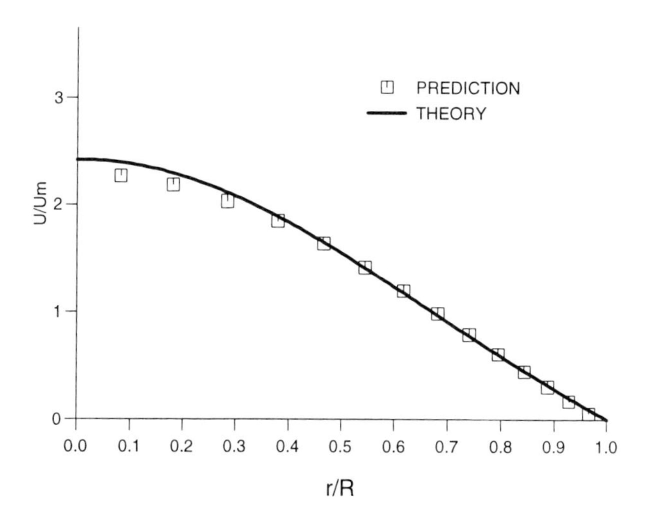

(a) Axial Velocity

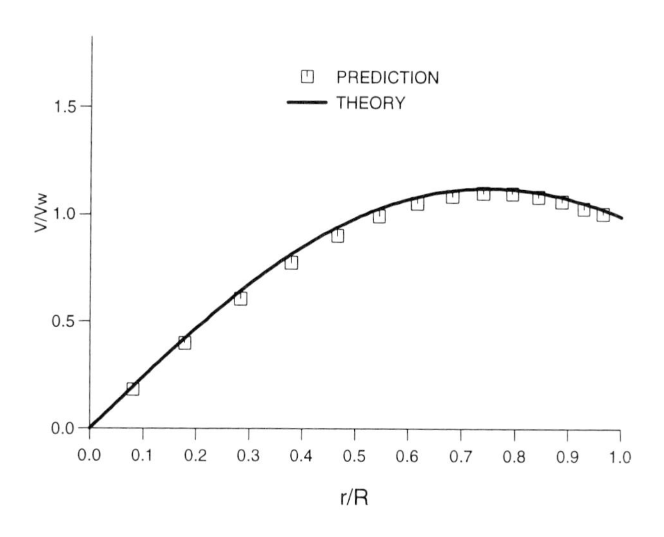

(b) Radial Velocity

Figure 13. Fluid-solid interaction case c, Table 3

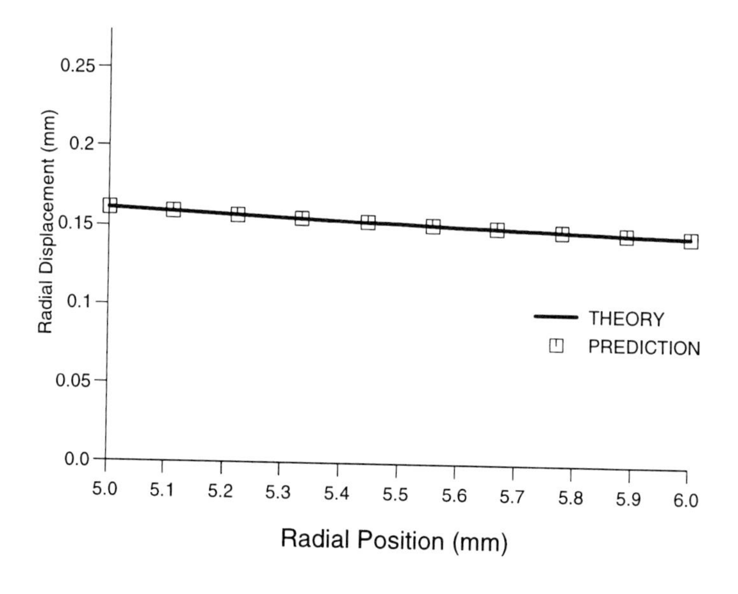

(a)

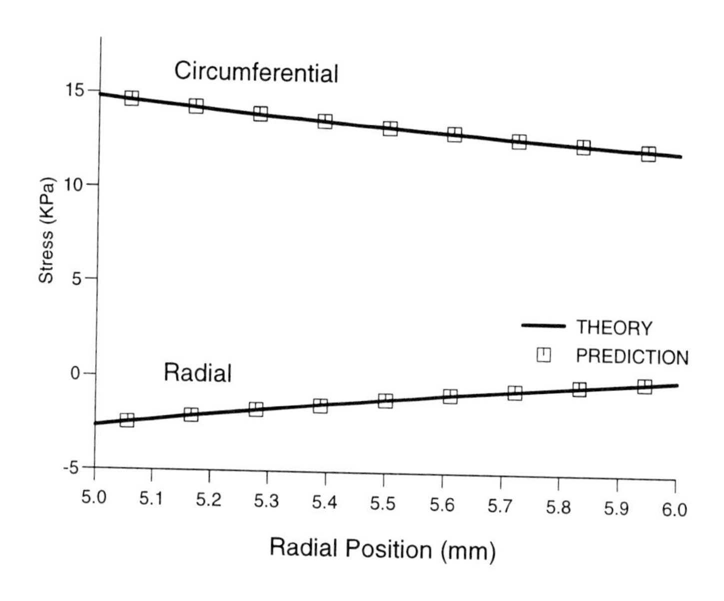

(b)

Figure 14. Fluid-solid interaction case d, Table 3

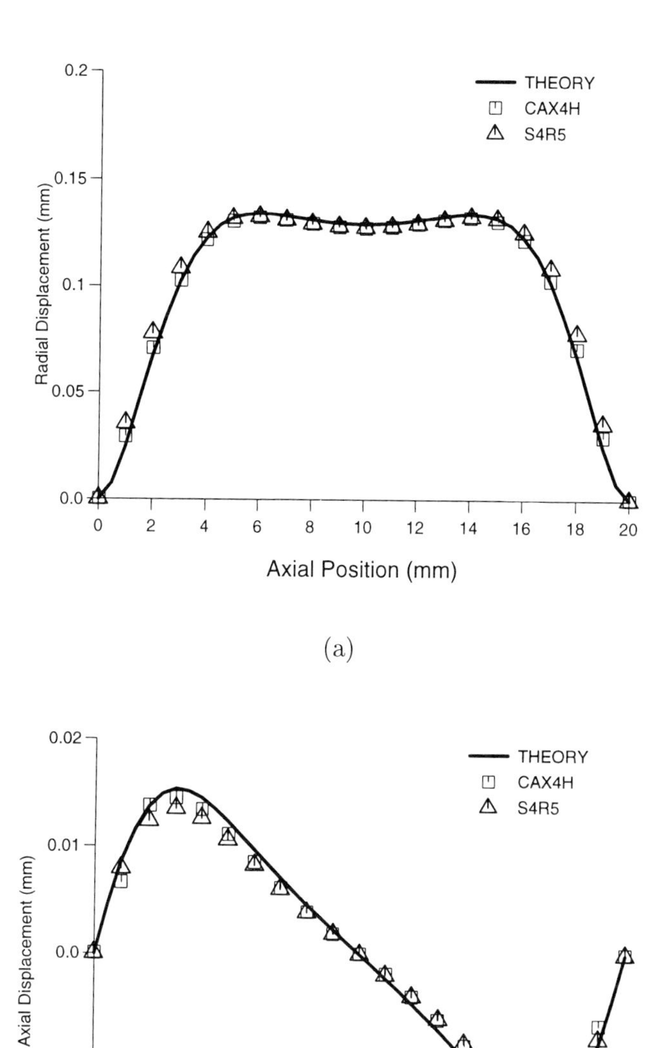

(a)

(b)

Figure 15. Fluid-solid interaction case e, Table 3

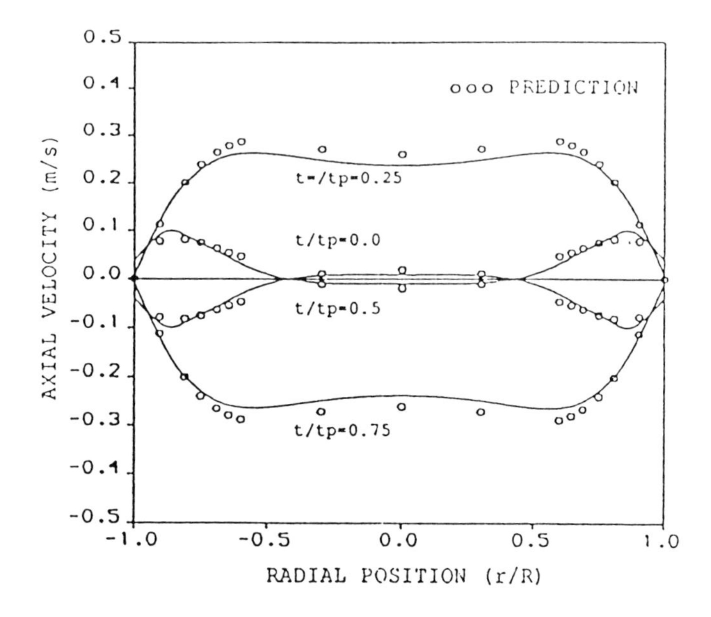

Figure 16. Fluid-solid interaction case f, Table 3

8 Visualisation of results-virtual reality

A numerical simulation, over a pulsatile cycle, will generate large amounts of data. The flow fields are three-dimensional and transient, the wall shear-stresses surface-following and transient also. The problem about visualisation of results arises because (a) the data are so large in number, and (b) the contorted nature of the vessels makes graphical display difficult. Figure 17 (taken from [17] for simulations of flow in a graft) illustrates the point (b) well. One approach is to use virtual reality methods, and in fact an industrial collaborator is doing this. Dr. Norman Rhodes of Mott MacDonald is applying in-house VR developed for conventional engineering [44] to physiological flows. Ultimately, we intend to extend the concept of integrated data processing/CFD software to include purpose-written VR post-processing. We are hoping to commence this project soon in the context of paediatric surgery.

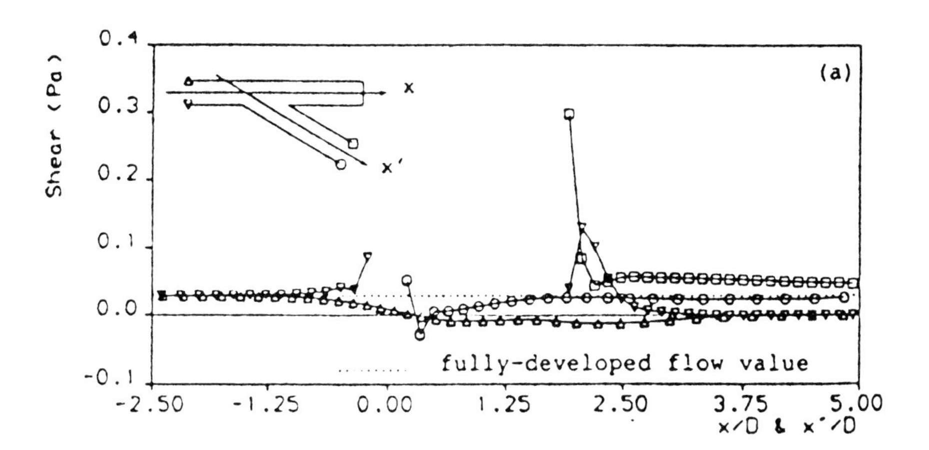

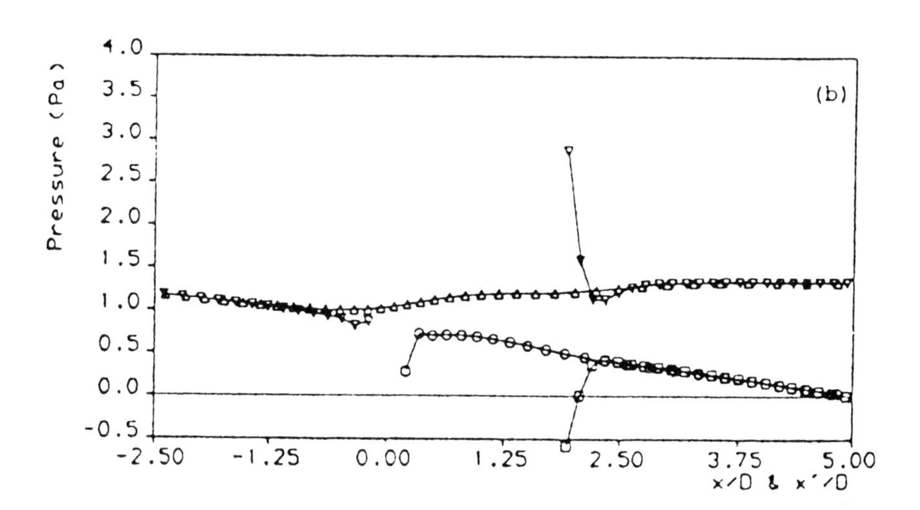

Figure 17. Visualisation difficulties of typical shear-stress patterns for graft

9 Conclusions

During the last decades numerical methods have been applied to study blood flow phenomena with great detail. Both to simulate accurately physiological conditions which are difficult to consider by experimental techniques, and to isolate different factors in an investigation, numerical methods are advantageous. In this paper, general discussion is made of relevance of CFD studies of blood flow for correlating its physiological state and pathological consequences. Furthermore, the combination of CFD and SM is of high promise in seeking to provide information both inside the flow domain and on the surrounding structures; the inclusion of wall compliance is a natural extension of our previous work, in the direction of greater fidelity. A detailed description of the way in which the combination of state-of-the-art CFD and SM can give reliable numerical simulations of haemodynamics in compliant human organs has been presented. The strategy is illustrated with typical results from a series of our attempts to model flow fields and subsequently to incorporate solid mechanics into haemodynamics in the cardiovascular system. A separate development has also been described for achieving complete image and data processing through to the numerical predictions.

Acknowledgements

We are grateful to British Heart Foundation, Welcome Trust and Engineering and Physics Sciences Research Council for sponsoring so much of our work.

Bibliography

1. Friedman, M.H. (1989) A biological plausible model of thickening of arterial intima under shear. *Atherosclerosis*, **9**, 511–522.

2. Lou, Z. and Yang, W.J. (1992) Biofluid dynamics at arterial bifurcations. *Critical Reviews in Biomed. Eng.*, **19**, 455–493.

3. Hofer, M., Rappitsch, G., Perktold, K., Trubel, W. and Schima, H. (1996) Numerical study of wall mechanics and fluid dynamics in end-to-side anastomoses and correlation to intimal hyperplasia. *J. Biomech.*, **29**, 1297–1308.

4. Perktold, K. and Rappitsch, G. (1995) Computer simulation of local blood flow and vessel mechanics in a compliant carotid artery bifurcation model. *J. Biomech.*, **28**, 845–856.

5. Liepsch, D. and Moravec, S. (1984) Pulsatile flow of non-Newtonian fluid in distensible models of human arteries. *Biorheol.*, **21**, 571–586.

6. Long, Q., Xu, X.Y., Collins, M.W., Bourne, M. and Griffith, T.M. (1998) Magnetic resonance image processing and structured grid generation of a human abdominal bifurcation. *Computer Methods and Programs in Biomedicine*. In press.

7. Taylor, T.W. and Yamaguchi, T. (1994) Three dimensional simulation of blood flow in an abdominal aortic aneurysm - steady and unsteady flow cases. *ASME J. Biomech. Eng.*, **116**, 89–97.

8. Perktold, K., Resch, M. and Florian, H. (1991) Pulsatile non-Newtonian flow characteristics in a three-dimensional human carotid bifurcation model. *ASME J. Biomech. Eng.*, **113**, 464–475.

9. Lonsdale, R. (1998) An algorithm for solving thermal-hydraulic equations in complex geometry. *The ASTEC code*, UKAEA Report.

10. Anonymous (1996) CFX 4: User Guide, Oxfordshire, UKAEA Technology Harwell Laboratory.

11. Anonymous (1991) User guide to FEAT. Engineering Analysis Centre, Nuclear Electric Plc.

12. Xu, X.Y. and Collins, M.W. (1990) A review of the numerical analysis of blood flow in arterial bifurcation. *Proc. Instn. Mech. Engrs. Part H: J. Eng. in Medicine*, **204**, 205–216.

13. Xu, X.Y. and Collins, M.W. (1994) Studies of blood flow in arterial bifurcation using a numerical simulation method. *Proc. Instn. Mech. Engrs. Part H: J. Eng. in Medicine*, **208**, 163–175.

14. Xu, X.Y. and Collins, M.W. (1995) Numerical modelling of blood flow in compliant arteries and arterial bifurcation. *Biofluid Mechanics*, Editor: H. Power, Computational Mechanics Publications, Southampton, Boston, 55–94.

15. Xu, X.Y., Collins, M.W. and Jones, C.J.H. (1997) A problem-oriented approach to the numerical modelling of haemodynamic problems. *Advances in Engineering Software*, **28**, 365–377.

16. Xu, X.Y., Collins, M.W. and Jones, C.J.H. (1992) Flow studies in canine artery bifurcations using a numerical simulation method. *ASME J. Biomech. Eng.*, **114**, 504–511.

17. Henry, F.S., Collins, M.W., Hughes, P.E. and How, T.V. (1996) Numerical investigation of steady flow in proximal and distal end-to-side anastomoses. *ASME J. Biomech. Eng.*, **118**, 302–311.

18. Henry, F.S., Shortland, A.P., Iudicello, F., Black, R.A., Jarvis, J.C., Collins, M.W. and Salmons, S. (1997) Flow in a simple model skeletal muscle ventricle: Comparison between numerical and physical simulations. *ASME J. Biomech. Eng.*, **119**, 13–19.

19. Henry, F.S. and Collins, M.W. (1993) A novel predictive model with compliance for arterial flows. *1993 Advances in Bioengineering, ASME BED,* Volume 26, 131–135.

20. Lan, T.H., Xu, X.Y., Hutton, A. and Collins, M.W. (1995) A numerical algorithm for solving coupled solid/fluid interaction problems. *Numerical Methods in Laminar and Turbulent Flows*, Editors: C. Taylor et al., Volume IX, Part 2, Pineridge Press, Swansea, 1539–1550.

21. Zhao, S.Z., Xu, X.Y. and Collins, M.W. (1997) A novel numerical method for analysis of fluid and solid coupling. *Numerical Methods in Laminar and Turbulent Flows*, Editor: C. Taylor, Volume 10, Pineridge Press, Swansea, 525–534.

22. Zhao, S.Z., Xu, X.Y. and Collins, M.W. (1998) The numerical analysis of fluid-solid interactions for blood flow in arterial structures. Part 2: Development of coupled fluid-solid algorithms. *Proc. Instn. Mech. Engrs. Part H: J. Eng. in Medicine*, **212**, 241–252.

23. Anayiotos, A. (1990) Fluid Dynamics at a compliant bifurcation model. PhD. Thesis, Georgia Institute of Technology.

24. Reuderink, P. (1991) Analysis of the flow in a 3D distensible model of the carotid artery bifurcation. Thesis, Eindhoven Institute of Technology, The Netherlands.

25. Pedley, T.J. (1980) The fluid mechanics of large blood vessels. Cambridge University Press.

26. Caro, C.G., Doorley, D.J., Tarnawski, M. et al. (1996) Nonplanar curvature and branching of arteries and non-planar-type flow. *Proceedings of the Royal Society of London*, **A, 452, (1944)**, 185–197.

27. Friedman, M.H., Deters, O.J., Mark, F.F., Bargeron, C.B. and Hutchins, G.M. (1983) Arterial geometry affects haemodynamics: A potential risk factor for atherosclerosis. *Atherosclerosis*, **46**, 225–231.

28. Long, Q., Xu, X.Y. and Collins, M.W. (1996) Generation of structure of the aortic bifurcation from magnetic resonance angiogram. *Proceedings of the First International Conference on Simulation Modelling in Bioengineering*, Editors: M. Cerrolaza, D. Jugo and C.A. Brebbia, Computational Mechanics Publications, 217–226.

29. Long, Q., Xu, X.Y., Collins, M.W., Griffith, T.M. and Bourne, M. (1997) Generation of CFD velocity boundary condition from cine MR phase-contrast images. *Proc. Medical Image Understanding and Analysis*, Oxford, 117–120.

30. Rohling, R., Gee, A. and Berman, L. (1997) Spatial compounding of 3-D ultrasound images. *Lecture Notes in Computer Science*, **1230**, 519–524.

31. Ciofalo, M., Hennessy, T.R. and Collins, M.W. (1996) Modelling nanoscale fluid and transport in physiological flows. *Medical Eng. and Physics*, **18**, 437–451.

32. Nerem, R.M. (1992) Vascular fluid mechanics, the arterial wall, and atherosclerosis. *ASME J. Biomech. Eng.*, **114**, 274–282.

33. Henry, F.S. and Collins, M.W. (1996) Numerical modelling of blood flow. *Advances in Hemodynamics and Hemorheology*, Editor: T.V. How, Volume 1, JA1 Press, 67–112.

34. Shortland, A.P. et al. (1996) Formation and travel of vortices in model ventricles: Application to the design of skeletal muscle ventricles. *ASME J. Biomech. Eng.*, **29**, 503–511.

35. Iudicello, F. et al. (1997) Comparison of haemodynamic structures in the skeletal muscle ventricle and in a human left ventricle. *Internal Medicine-Clinical and Laboratory*, **5**, 1–10.

36. Xu, X.Y. and Collins, M.W. (1996) Fluid dynamics in stents. *Endoluminal Stenting*, Editor: U. Sigwart, W.B. Saunders, 52–59.

37. Zhao, S.Z., Xu, X.Y. and Collins, M.W. (1998) The numerical analysis of fluid-solid interactions for blood flow in arterial structures. Part 1: A review of models for arterial wall behaviour. *Proc. Instn. Mech. Engrs. Part H: J. Eng. in Medicine*, **212**, 229–240.

38. Ciofalo, M., Collins, M.W. and Hennessy, T. (1998) A review of nanoscale fluid dynamics in physiological processes. *Computational Mechanics Pubs.* In press.

39. Long, Q., Xu, X.Y., Collins, M.W. et al. (1998) The combination of magnetic resonance angiography and computational fluid dynamics: A critical review. *Critical Reviews in Biomed. Eng.* In press.

40. Collins, M.W. and Xu, X.Y. (1990) A predictive scheme for flow in arterial bifurcations: Comparison with laboratory measurements. *Proc. NATO Workshop on Biomech Transport Processes, Corsica, 1989, Biomechanical Transport Processes*, Editors: F. Mosora et al., Plenum Press, 125–133.

41. Iudicello, F. et al. (1996) Numerical simulation of the flow in simple model skeletal muscle ventricles. *Harwell CFX Conf. Proceedings.*

42. Henry, F.S. and Collins, M.W. (1998) Numerical predictions of flow in a model compliant artery. Submitted to *J. Fluids and Structures.*

43. Long, Q. et al. (1998) Reproducibility analysis of 3D geometrical reconstruction of a human arterial bifurcation. *British Second Medical Image Processing and Analysis Procs.*, Oxford University Press.

44. Rhodes, N. (1998) Application of virtual reality to the visualization of complex flows. *Int. Conf. on Optical Methods and Data Processing in Heat and Fluid Flow, I.Mech.E. Proc.*, 291–299.

Experimental and Computational Modelling of Flow through an Arterial Bypass Graft

C.J. Bates, D. Williams and D.M. O'Doherty

Division of Mechanical Engineering and Energy Studies, Cardiff University

Abstract

Distorted mean velocity and turbulence intensity profiles in the horizontal plane of a 30° Y junction at various Reynolds numbers show the significance of the graft geometry and flow conditions on the localisation of atherosclerosis and intimal hyperplasia. Fluctuating wall shear stresses in the graft, which are associated with disturbed flow, are believed to be important factors in the development and localisation of intimal hyperplasia.

1 Introduction

Atherosclerosis is a common disease in men and women over 60 years old. It is a disease that commences early in life. A study of American soldiers killed in action in Korea whose average age was 22.1 years old revealed that 77% of the hearts autopsied showed gross evidence of coronary atherosclerosis. Atherosclerosis results from the build up of plaques or fatty deposits, which can cause a critical reduction in the cross sectional area of the blood vessels. The result is a reduction in the circulation of blood to a symptomatic level. Symptoms of the disease include pain and ischemia. One example of this disease is plaque build up in the carotid arteries that reduces the flow of blood to the brain and eventually leads to a stroke. Similarly, obstruction of the blood flow through the coronary arteries can result in a heart attack.

The surgical treatment of atherosclerosis includes angioplasty and arterial bypass grafts. Angioplasty is not a suitable treatment if there are more than three occluded vessels. In such patients, the surgeon can perform bypass surgery to improve the flow of blood through the artery. At present arterial bypass graft is a routine and effective re-constructive vascular surgery treating occlusive arterial disease.

The bypass vessel can be the patients saphenous vein or may be constructed from a synthetic material such as polytetrafluoroethylene (PTFE). The cumulative patency rate for saphenous vein grafts to isolated popliteal segments is 70% for five years, whereas the synthetic grafts have only a 40% five years patency rate. This rate can be improved for the PTFE grafts, if it fails in the first 20 months, by performing thrombectomies and graft extensions [1]. These procedures are possible because PTFE is highly durable during re-operation of a failed bypass. By contrast saphenous vein grafts usually require a full replacement when they fail.

Intimal thickening (hyperplasia) is the main cause of failure in these bypass grafts [2]. Although intimal thickening is a normal feature of the healing response of arteries and of artery adaptation to flow conditions, it may continue after normal post-operative healing and thus narrow the artery lumen. This leads to the reduction or complete blockage of blood flow through the artery [3].

Local fluid dynamic factors have been shown to be closely related to the localisation of intimal thickening and atherosclerosis [4]. The research done on the bifurcations in the human body indicates that the disease develops largely in regions of relatively low and fluctuating wall shear stresses, flow separation and departure from unidirectional flow. The research suggests that the disease does not develop in regions of moderate to high wall shear stresses.

The role of fluid mechanics in atherogenesis has been investigated for many years. It is still not clear as to the causes for the initiation and progression of the arterial lesions. Direct evidence of haemodynamic involvement is lacking. The only indirect evidence is that atherosclerotic lesions do not occur randomly in the arterial vasculture but tend to be localised at sites where blood flow effects differ from those in unstenosed straight arteries. These sites include sharp bends and bifurcations [5].

Two popular fluid dynamics theories for the disease are based on:

- High shear stress which can damage the endothelial lining and, therefore, initiate the formation of plaque;
- Low shear stress, which presents a favourable condition to initiate atherosclerosis by shear dependant mass transfer of blood constituents.

Research, over the years, into this subject has considered many different configurations of the arterial bypass graft in an attempt to design the optimal graft that completely eliminates the build up of intimal hyperplasia. Many designs have been constructed and tested for various artery diameters, flexible walls, hood lengths, anastomotic branching angles, proximal outflow segment (POS): distal outflow segment (DOS) ratios and flow rates [6–8]. The precise geometry of the graft is a key factor that influences local haemodynamics. To date the optimal design has been very elusive.

From a literature review it is evident that most variables in the design of the arterial bypass graft have been examined thoroughly. This paper will consider what effect the fillet radius at the intersection between the host artery and the graft will have on the flow behaviour through the bypass. Prior to the current investigation this variable has not been subjected to detailed research. Details of a new 50 mm diameter bifurcation glass flow facility will be presented, this facility has been designed to cover pulsatile flow over a mean Reynolds number range from 200 to 10000. This range, which covers that found in the human body, will be used as well as proximal to distal outlet segment flow ratios.

Mean velocity and root mean square turbulence profiles are used as an indicator to disturbed flow or non-uniform haemodynamic flow factors; this may then be the best predictor for locating sites where intimal hyperplasia and atherosclerosis are likely to develop. By monitoring the disturbed velocity and turbulence intensity profiles in the 5 mm fillet radius graft model for different flow conditions it has been possible to establish how the DOS: POS flow ratios and the Reynolds number influences the flow in the junction.

2 Experimental facility and apparatus

The experimental facility used in the present study provides a constant head to flow through a 30° Y junction, which is split into various distal to proximal outlet flow ratios. Water was pumped from a low-level reservoir to the constant head tank, which incorporated an overflow facility, the excess water being returned to the reservoir as shown in Figure 1. From the constant head tank the water passes through a Watson-Marlow 704 U high flow peristaltic pump, via a length of marprene tubing, into a 5 m length of QVF transparent glass pipeline with an internal diameter of 50 mm and 4 mm wall thickness. This peristaltic pump supplies the rig with pulsatile flow at Reynolds numbers of 500, 2000 and 10000. The 5 m length of pipeline is inclined at an angle of 30° to the horizontal, with a length to diameter ratio of 100 fully developed flow profiles are achieved upstream of the entry into the 30° Y junction, as shown in Figure 3, for Reynolds numbers of 10000 and 500 and various DOS: POS flow ratios.

Six different 30° Y junctions were designed, built and are to be examined, these models were made of glass pipe to provide optical access for a conventional one dimensional forward scatter laser Doppler anemometry system. These junctions differ from each other only by the radius of the fillet between the graft and the host artery; the radii of the six fillets are 0, 5, 10, 15, 20 and 25 mm. The objective of the overall study being to establish how the radius of the fillet affects the flow conditions inside the graft model.

After the Y junction the rig had glass pipelines of the same diameter, each 3 m length, in both the distal and proximal legs of the junction. Two butterfly valves were installed at the end of these legs to enable accurate control of the distal and proximal outlet segment flows. For this experiment three different distal to proximal outlet segment flow ratios were chosen 80:20, 60:40 and 40:60.

The LDA system used comprised of a Dantec 60X17 optical probe with 8 mm beam spacing and a 50 mm focal length lens: this probe was powered by a 5 W argon-ion laser (Coherent Innova 90) via a Dantec 60X40 transmitter box. Directional ambiguity was removed through the use of Bragg cell frequency shifting at 40 MHz. The receiving optic comprising of a photomultiplier, pinhole and the appropriate filter for the green wavelength (514.5 nm) was used in the forward scatter light collection mode. This set-up was further improved by the use of a close up lens with an 80 mm focal length, which increased its focussing capability.

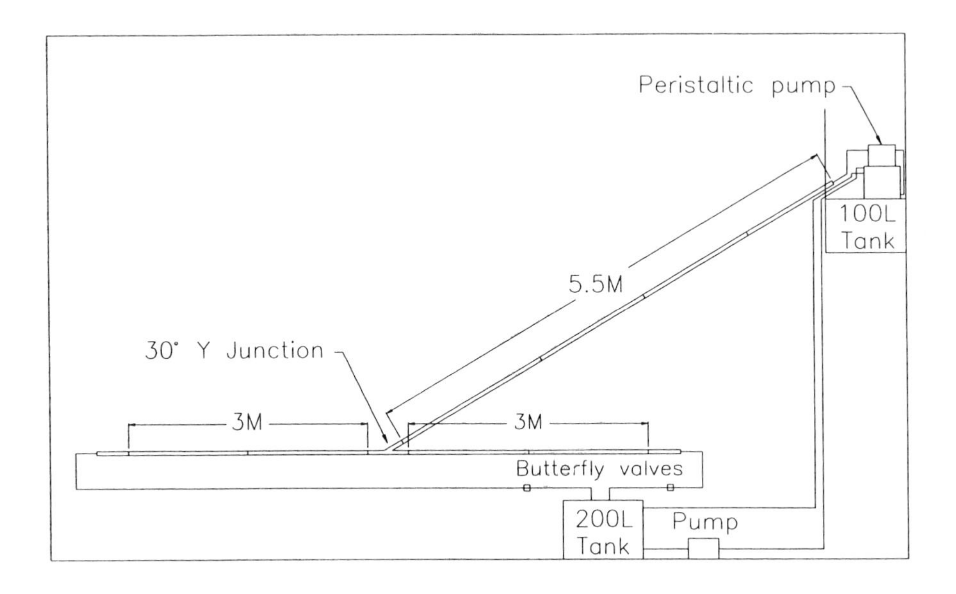

Figure 1. The experimental facility

The signal from this set-up was processed using a Dantec 57N20 enhanced Burst Spectrum Analyser processor and a software package created by Dantec called Burstware. The laser Doppler anemometry probe was mounted on a three dimensional traverse controlled by a Dantec 57G15 traversing interface which was then monitored and operated via a computer. This allowed measurements of displacements to a resolution of 0.01 mm with an error of approximately 3.6%. Data analysis was based upon 15000 validated Doppler bursts, data rates up to 20 kHz were experienced.

Ground mica (Timeron) was used for seeding the flow. The expected velocity error due to the inability of the seed particles to follow the fluid oscillation both within and outside any re-circulation regions was confirmed to be less than 0.5% [9].

3 Results

The results presented in this paper are only a sample of the Reynolds numbers measured with the laser Doppler anemometry system. These graphs show the horizontal plane along the centre line of the host artery. The diagram, Figure 2, is a vertical cross section of the graft model. The profiles on the graphs, Figures 3(b)–3(m), are all 20 mm from each other and their positions on the graph correspond to their position along line A-A on Figure 2. All graphs

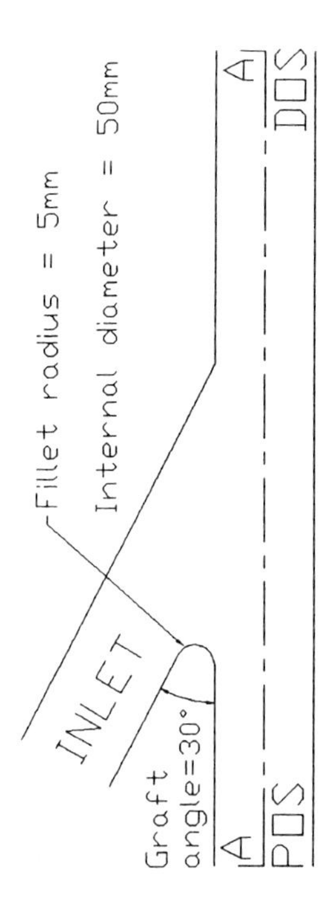

Figure 2. Vertical cross-section of the 30° Y junction

have been plotted on the same scale axes. The various flow configurations presented are DOS: POS flow ratios of 80:20, 60:40 and 40:60 and Reynolds numbers of 500 and 10000.

The graphs, Figures 3(b)–3(m), allow a comparison to be made between the velocity and turbulence intensity profiles in the graft model at the chosen Reynolds numbers. The turbulence intensity graphs show the root mean square turbulence values divided by the mean velocity at that point. When the turbulence intensity values are large the RMS turbulence values are of the same order or larger than the mean velocity. This suggests that the fluctuations in the measured instantaneous velocity are as large as the mean velocity.

The velocity profile graphs for both Reynolds numbers and a DOS flow of 80%, Figures 3(b) and 3(c), show some very disturbed velocity profiles, but also some similarities in the structure of the flow inside the junction for both Reynolds numbers. The lower velocities seen near where the graft enters the host artery are thought to be caused by the flow swirling in and out of the horizontal plane. The one-dimensional laser Doppler system only measures the axial component of the velocity, not the true velocity.

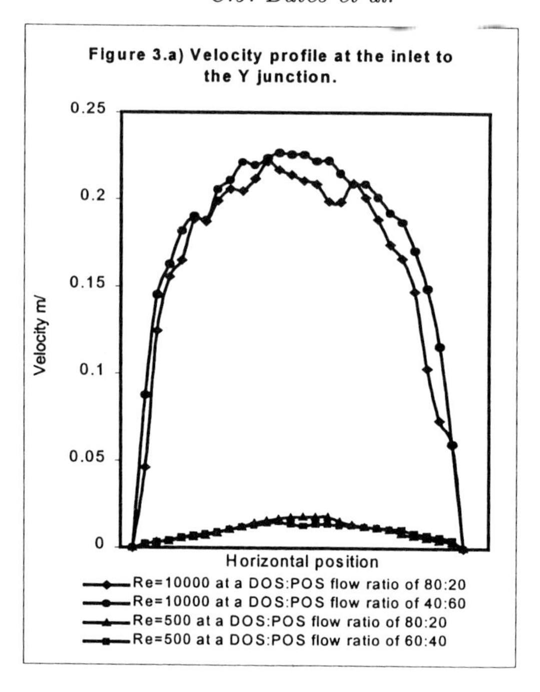

Figure 3. (a)

As the flow follows the graft wall, on either side of the junction, to the bed of the host artery, it meets the flow from the other side of the junction creating a swirl on both sides of the pipe. This swirl is then superimposed onto the axial flow in the pipe resulting in a large helical swirl along the pipeline. The turbulence intensity graphs for these flow conditions, Figures 3(d) and 3(e), show that the flow in the model with the Reynolds number of 500, Figure 3(e), is less disturbed/turbulent than the model with the higher Reynolds number, Figure 3(d). Both models show that the turbulence decays considerably as it nears the outlets.

The larger areas of turbulence intensity are the areas where fluctuations in the shear stresses are likely to be found. Fluctuating shear stresses are believed to be a factor that promotes the development of intimal hyperplasia leading to graft failure.

The velocity profiles in the junction for the DOS flow ratio of 60%; Figures 3(f) and 3(g) show much more disturbance than the previous flow conditions. The number of points on the turbulence intensity graphs, Figures 3(h) and 3(i), showing more signs of turbulence have increased showing many more flow disturbances and large fluctuations in the measured velocities. The graph for the Reynolds number of 500, Figure 3(i), shows fewer disturbances than the model with a Reynolds number of 10000, Figure 3(h).

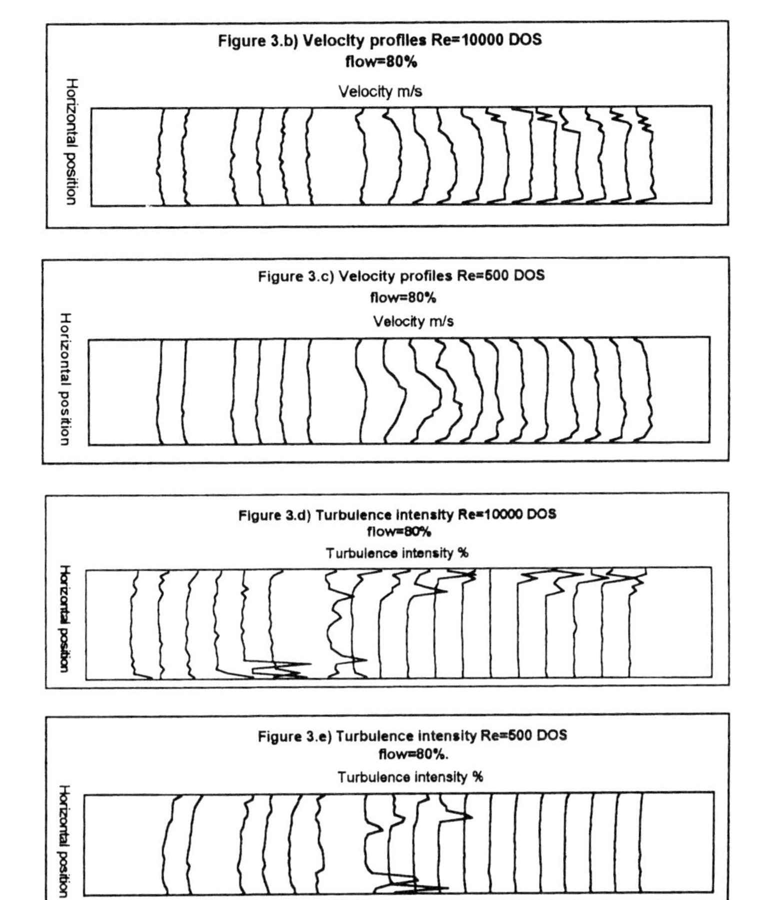

Figure 3. (b), (c), (d) and (e)

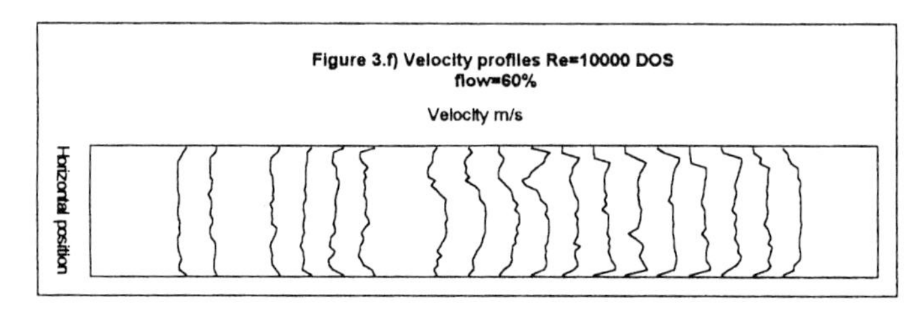

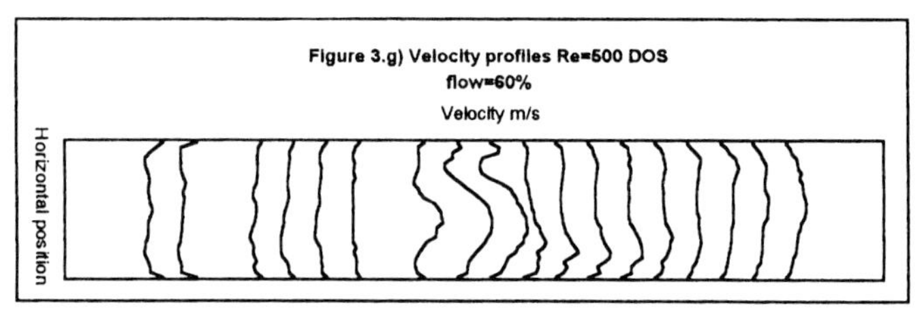

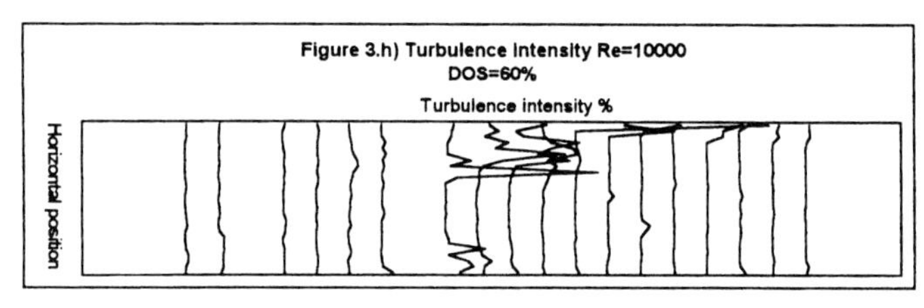

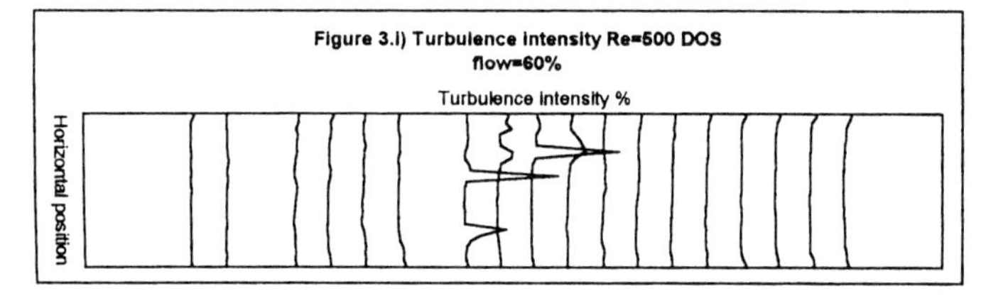

Figure 3. (f), (g), (h) and (i)

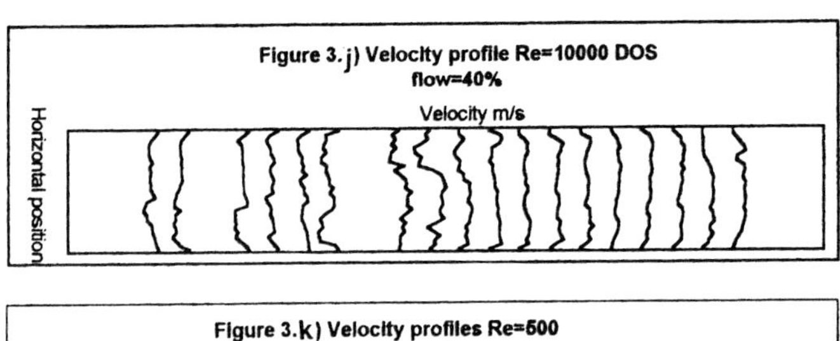

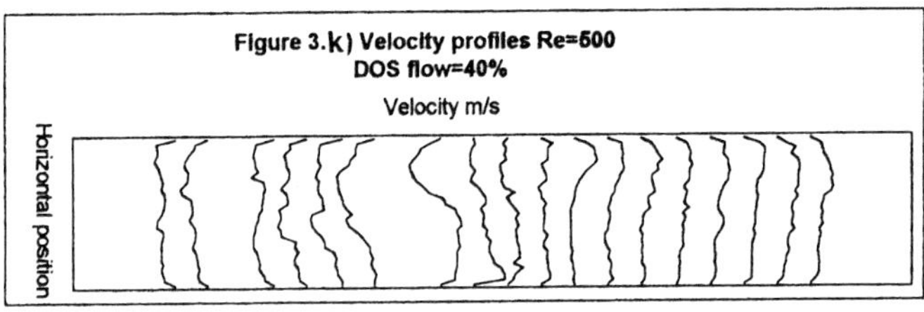

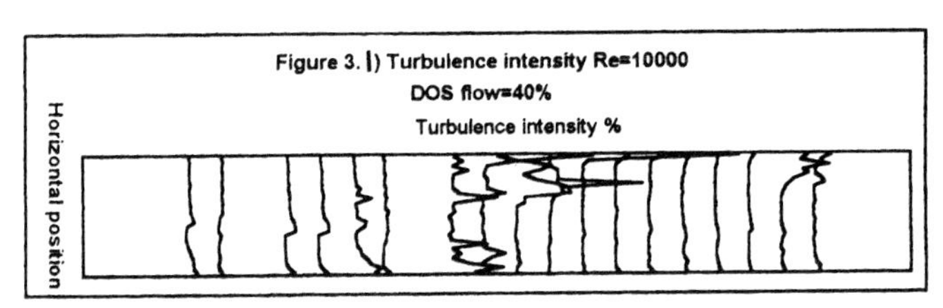

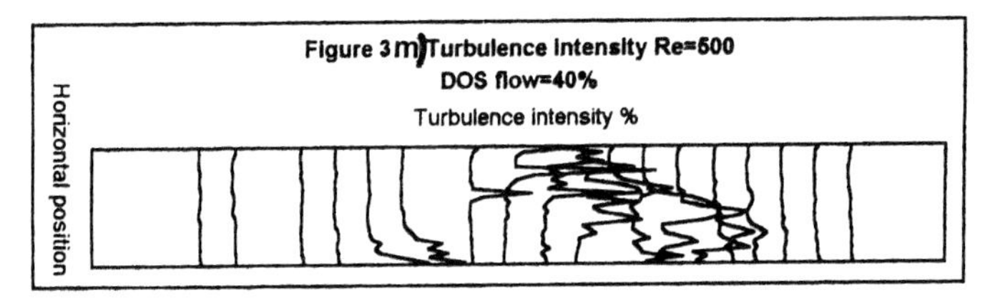

Figure 3. (j), (k), (l) and (m)

The final set of graphs for a DOS: POS flow ratio of 40:60 for the two Reynolds numbers, Figures 3(j) and 3(k), show even more flow disturbances. The velocity profiles toward the proximal outlet are more distorted than the previous flow configurations and it also takes much longer for the flow to become less chaotic. The turbulence intensity graphs, Figures 3(l) and 3(m), for both Reynolds numbers show that the flow is very disturbed. In this case the lower Reynolds number shows more evidence of turbulence than the higher.

4 Discussion

In all flow configurations some redevelopment of the flow occurs as it nears the two outlets of the graft model, which are 220 mm and 140 mm respectively from the centre of the model where the graft enters. The flow develops more rapidly in the models with the majority of the flow leaving the graft via the distal outlet segment i.e. DOS: POS of 80:20 and 60:40. The model with the flow ratio of 40:60 takes much longer for the flow to become less chaotic which suggests that the flow is very disturbed and has a great deal of kinetic energy especially toward the proximal outlet. The velocity profiles inside the junction for both Reynolds numbers show that the flow is very disturbed compared to the fully developed parabolic profiles seen at the inlet, Figure 3(a). The flow patterns observed in the junction are very similar to that seen by Keynton et al. [7].

Considerable decay in the turbulence intensity values is also observed as it nears the outlets in all but the 40:60 model. Decay in these values suggests that the flow instabilities are becoming less pronounced and that the flow is redeveloping. In the 40:60 model the turbulence intensity values remain high toward the proximal outlet confirming the belief that the flow contains much instability. The results show evidence that there are instabilities in the flow that may exist for a considerable distance downstream. These disturbances, which may be time dependent, are unwelcome in bypass grafts as mentioned earlier because they cause fluctuating shear stresses that lead to the development of intimal thickening and graft failure.

5 Conclusion

The measurements confirm that the 30° Y junction causes instabilities in the flow which may exist a considerable distance downstream. The high turbulence intensity values seen at some points in the junction confirm that the instantaneous velocity of the flow fluctuates considerably. The fluctuations in the velocity suggest that there are fluctuating wall shear stresses in the junction. It is believed that these fluctuating shear stresses located around where the graft enters the host artery, is the main cause of graft failure. It has been observed that these disturbances are greatly exaggerated as the DOS flow decreases and more flow leaves the graft via the proximal outlet segment. It was also noted that the lower the Reynolds number then the less chance there is of intimal thickening, this is because there is less turbulence at the lower Reynolds number.

Bibliography

1. Veith, F.J., Gupta, S., and Daly, V. (1980). Management of early and late thrombosis of expanded polytetrafluoroethylene femoro popliteal bypass graft: favourable prognosis with appropriate re-operation. *Surgery 87*, **5**, p. 581.

2. Echave, V., Koornich, A.R., and Haimov, M. (1979). Intimal hyperplasia as a complication of the use of the polytetrafluoroethylene graft for femoropopliteal bypass. *Surgery 86*, **6**, p. 791.

3. Logerfo, F.W., Quist, W.C., Nowak, M.D., Crawshaw, H.M., and Haudenschild, C.C. (1983). Downstream anastomotic hyperplasia. *Surgery 197*, **4**, p. 479.

4. Ojha, M. (1987). An experimental investigation of pulsatile flow through modelled arterial stenoses. *Ph.D. Thesis*, University of Toronto.

5. Bassiouny, H.S., White, S., Glagov, S., Choi, E., Giddens, D.P., and Zarins, C.K. (1992). Anastomotic intimal hyperplasia: mechanical injury or flow induced. *Journal of Vascular Surgery*, **15**, pp. 708–717.

6. Fei, D.Y., Thomas, J.D., and Rittgers, S.E. (1994). The effect of angle and flow rate upon haemodynamics in distal vascular graft anastomoses: a numerical model study. *Journal of Biomechanical Engineering*, **116**, pp. 331–336.

7. Keynton, R.S., Rittgers, S.E., and Shu, M.C. (1991). The effect of angle and flow rate upon haemodynamics in distal vascular graft anastomoses: an in vitro model study. *Journal of Biomechanical Engineering*, **113**, pp. 458–463.

8. White, S.S., Zarins, C.K., Giddens, D.P., Bassiouny, H.S., Loth, F., Jones, S.A., and Glagov, S. (1993). Haemodynamic patterns in two flow models of end to side vascular graft anastomoses: effect of pulsatility, flow division, Reynolds number and hood length. *Journal of Biomechanial Engineering*, **115**, pp. 104–111.

9. Sierra-Espinosa, F.Z. (1997). The turbulence structure of the flow in a 90° pipe junction: a comparison of numerical predictions to experimental laser doppler and particle image velocimetry results. *Ph.D. Thesis*, University of Wales, Cardiff.

The Convergence Technique in Numerical Solutions of Coupled Fluid-Wall Problems

S.Z. Zhao*, **X.Y. Xu***, **M.W. Collins****, **A.V. Stanton**†, **A.D. Hughes**† **and S.A. Thom**†

**Department of Chemical Engineering and Chemical Technology, Imperial College, London, **School of Engineering Systems and Design, South Bank University, London, and †Department of Clinical Pharmacology, Imperial College School of Medicine at St. Mary's, London*

Abstract

A comprehensive coupled fluid-wall interaction dynamics model was developed for the simulation of blood flow in arteries. Blood flow was treated as pulsatile, laminar, Newtonian and incompressible. The structural model accounted for both material and geometric nonlinearities. A multiblock body fitted grid was used to subdivide the flow domain into computational finite volume cells. Shell elements were used to discretize the arterial wall. A finite volume computational fluid dynamics code and a finite element structural dynamics code were used to solve the flow and structure equations, respectively. This paper discusses some important issues related to the coupled model, particularly numerical convergence. In addition to convergence considerations within each code, convergence between the two codes must also be dealt with. And a relaxation technique has been introduced to improve and accelerate the convergence performance.

1 Introduction

Fluid-wall interaction problems exist in the cardiovascular system where blood interacts dynamically with its surroundings. In order to understand both the haemodynamics and vessel wall mechanics in a real artery, computer simulation of this complex system has become more and more demanding [1, 2]. In the present paper, a numerical solution procedure for solving general fluid-wall interactions is developed. In this approach, the time-dependent Navier-Stokes equations of the fluid are not explicitly assembled with the dynamics equilibrium equations of the wall, but instead an iterative procedure is used to obtain the solution. Combining two established commercial codes, CFX for fluid flow and ABAQUS for wall motion [3, 4], this iterative algorithm preserves the nature of software modularity and offers a distinct advantage in that the most efficient codes for fluid flow and structural analysis can be used to deal with complex problems.

A number of numerical experiments have been performed to test the convergence of the iterative procedure implemented. In some cases it is necessary to introduce a relaxation factor because the fluid domain boundary is likely to oscillate if the total geometric deformation as a result of the structural analysis is fed in as the new boundary. It was found that by updating the deformations incrementally, the dynamic loads resulting from the flowing fluid would not change drastically. This relaxation method assimilates a portion of the geometric deformation that are relaxed from their full values at each update interval. Thus the stable state can be reached under a relaxation factor. The proposed scheme can also be used for solving a wide range of engineering problems involving fluid/solid interactions.

Numerical results are presented to demonstrate the convergence characteristics of the proposed scheme.

2 The coupled scheme

Numerical methods applied to coupled problems lead to the solution of a set of nonlinear algebraic equations [5]. Thus, the choices for solving a coupled problem are twofold:

- Strategy 1: to treat all the domains simultaneously. This results in a single set of algebraic equations involving all the relevant variables. In general, these variables are not homogeneous, as they represent discretization of different domains and/or different physical phenomena.
- Strategy 2: to treat the domains one at a time sequentially, considering the coupling terms as forcing terms on the right-hand side of the equations. This leads to several sets of algebraic equations (one per domain), each of them to be solved for variables related to one domain, but with the right-hand side depending on variables related to the other domains.

Strategy 1 requires the development of a special-purpose code, probably involving collaboration among different areas of expertise. Standard engineering software developed for uncoupled problems may be of little use when writing such a program, owing to its particular structure. The outcome of this may well be a complicated code, difficult to maintain, modify or upgrade, and even difficult to use. However, the advantage is that instant convergence of the numerical solution becomes feasible.

Strategy 2 on the other hand, allows each domain/problem to be tackled on its own. The codes used may be either new or existing programs, appropriately modified to account for the coupling terms. Each of these codes may be developed by an expert or a team of experts in the particular field, using optimal strategies for each of them. The outcome of this should be a set of computer programs, highly efficient on their own, easy to maintain, modify or upgrade, each independent of the others.

Based on the advantages and disadvantages of the two approaches, we developed a novel numerical method for coupled solid/fluid problems by combining two well established commercial codes. Compared to the direct solution approach, this iterative scheme preserves the nature of both finite volume and finite element approaches. The equations for the fluid motion and solid deformation are solved separately and then coupled externally in an iterative manner at a cycle level. Employing the user-defined subroutines in both codes, the updated pressure and deformation distributions at each time step can be transferred between CFX4.1 and ABAQUS on an interface mesh. Details of the coupled algorithm can be found in Zhao et al. [6].

A convergence criterion has to be introduced to control the iterative procedure. This is based on the difference between displacements at two successive iterative cycles at each time step(DD). Under this criterion, the convergence is achieved when there is little or no change in the position of interface surface at each time step. A similar convergence criterion has been adopted by Müller and Jacob [7], while Hofer et al. [8] employed a different criterion based on the difference in successive velocities.

3 Applications of the coupled method

Numerical predictions were carried out for pulsatile flows in three compliant models of human carotid artery bifurcations (Figure 1) to demonstrate the validity of the coupling algorithm developed.

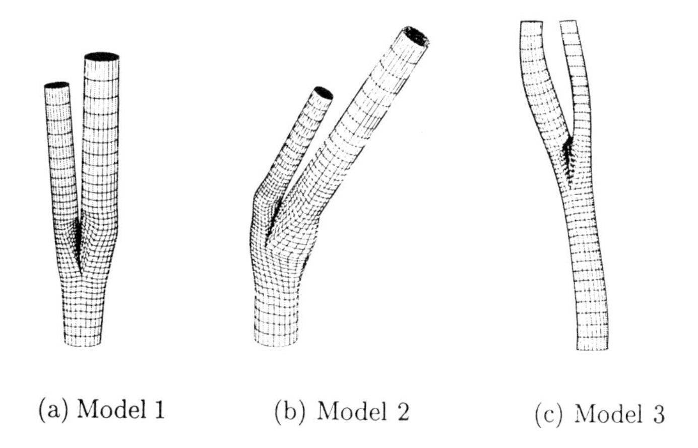

(a) Model 1 (b) Model 2 (c) Model 3

Figure 1. Geometries of the three carotid bifurcation models

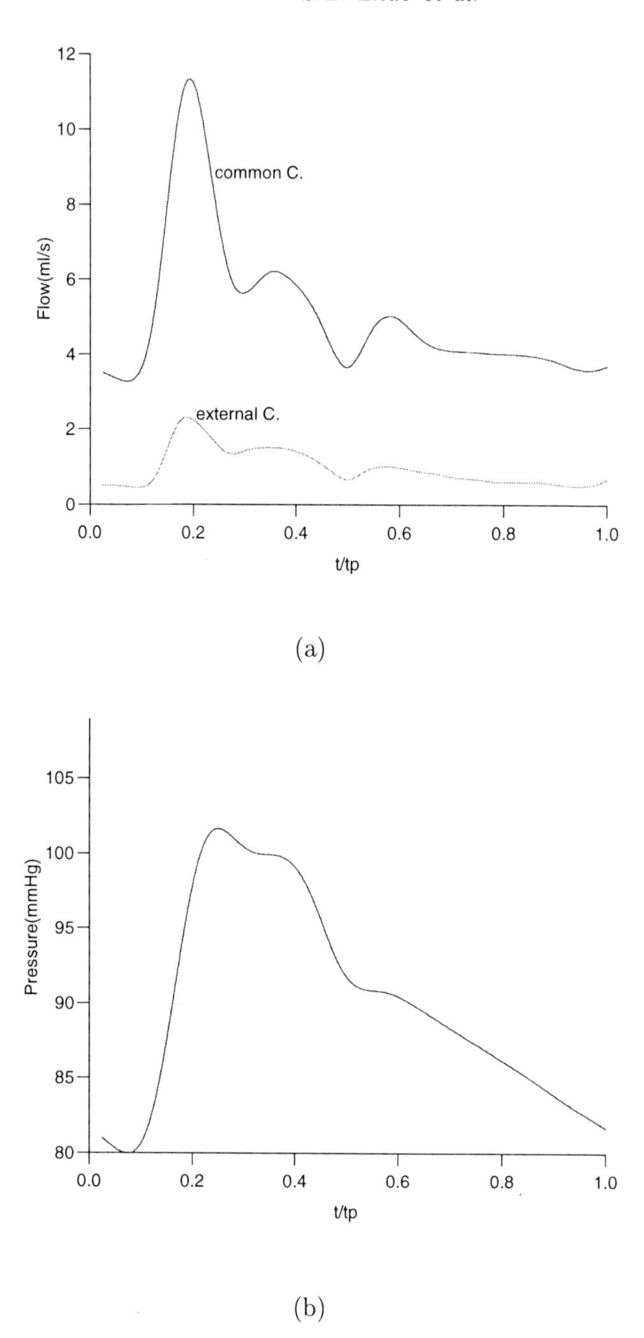

Figure 2. Boundary conditions for the calculations. (a) Flow pulse waveforms in the common and external carotid arteries, and (b) Pressure pulse waveform in the internal carotid outlet

The configurations are defined as the diastolic reference geometry. The essential geometric parameters are as follows:

- Model 1: common carotid diameter is 6.25 mm, maximum sinus diameter is 6.0 mm, internal carotid diameter is 6.0 mm, external diameter is 4.3 mm, angle between internal and external axes is 5°;
- Model 2: common carotid diameter is 6.0 mm, maximum sinus diameter is 6.2 mm, internal carotid diameter is 4.3 mm, external diameter is 3.3 mm, angle between the common and internal axes is 30° and angle between the common and external axes is 25°;
- Model 3: common carotid diameter is 5.9 mm, internal carotid diameter is 4.6 mm, external diameter is 3.2 mm.

As the last model was reconstructed from *in vivo* MR angiogram, there is no symmetric plane as in the first two cases.

The nonlinear behaviour of the artery wall has been approximated, by assuming it to be incrementally linearly elastic over the pressure range from the diastolic level to systolic phase [9]. The incremental elastic modulus of the arterial wall is 5×10^5 Pa. The thickness (h) of the carotid artery walls is 8–10% of the vessel diameters which are 0.5 mm, 0.5 mm, and 0.45 mm respectively for the three models. To prevent rigid body motion, the ends of the model were constrained against motion. The flow domain discretizations result in 6480, 6480, and 10240 eight-noded brick elements and the wall discretizations yield 864, 864, and 1280 four-noded shell elements respectively for the three models. Flow waveforms obtained from Perktold and Rappitsch [9] were applied as the time-dependent boundary conditions. The flow pulse waveform at the common carotid inlet and the external carotid outlet as well as the pressure pulse waveform at the internal carotid outflow boundary are shown in Figure 2. As the boundary conditions used do not correspond to the individual models studied here, they are meant to demonstrate the coupled method and convergence technique rather than to simulate the real cases.

4 Convergence technique

The coupling for a fluid/solid problem is not a trivial task and several difficult issues related to both theory and implementation aspects may arise. Whenever there is iteration, there is always convergence problem. The problem of convergence within the calculations of each single code lies in the scope of the code itself. However, the convergence behaviour between the codes has to be monitored and controlled for a coupled solution. A summary of the convergence behaviour for the three coupled models is given in Table 1.

Table 1. Summary of the convergence behaviour for the three coupled models

Iteration	1-2	2-3	3-4	4-5	5-6	6-7
MODEL 1 DD(%)	68.3	14.9	4.2	1.9	1.1	
MODEL 2 DD(%)	187.89	172.79	46.96	12.88	4.64	2.33
MODEL 3 DD(%)	317.22	221.38	174.57	152.44	124.31	103.04

Table 2. Convergence behaviour of Model 3 after relaxation

Iteration	1-2	2-3	3-4	4-5	5-6	6-7	7-8
DD(%)	159.63	48.14	27.19	12.38	6.63	4.44	2.61

Table 3. Convergence behaviour of Model 1 with different grid density and time steps

Iteration	1-2	2-3	3-4	4-5	5-6	6-7
MODEL 1(a) DD(%)	68.32	14.99	4.21	1.92	1.03	
MODEL 1(b) DD(%)	102.0	128.0	189.7	399.3	894.4	1415.6

Table 4. Convergence behaviour of Model 1 after time step refinement without and with a relaxation factor of 0.5

Iteration	1-2	2-3	3-4	4-5	5-6	6-7	7-8
MODEL 1(b) before relaxation DD(%)	102.0	128.0	189.7	399.3	894.4	1415.6	
MODEL 1(b) after relaxation DD(%)	79.36	32.53	20.49	12.07	6.83	4.22	1.94

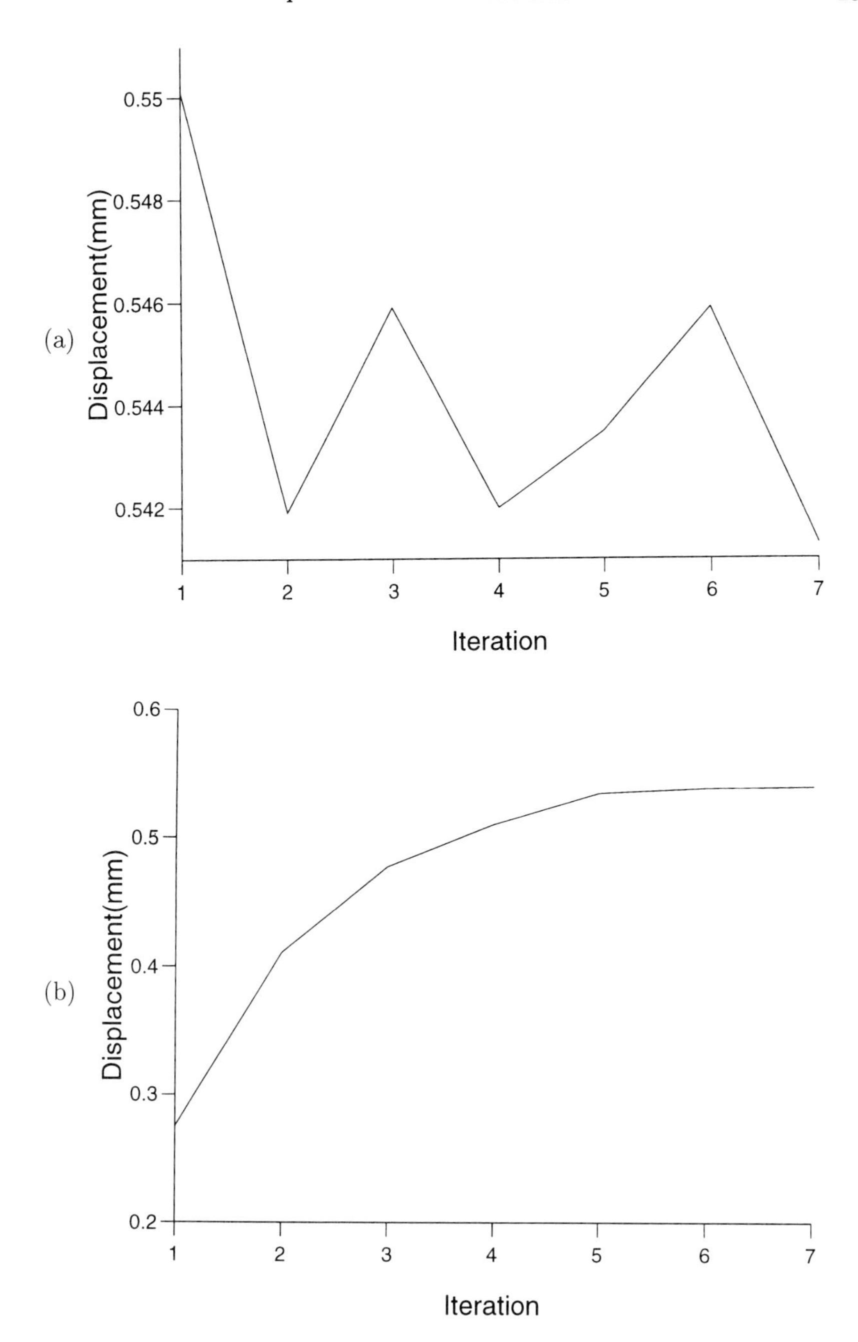

Figure 3. Oscillating geometry before relaxation and converging geometry with a relaxation factor of 0.5 at the maximum displacement point of Model 1

The calculations showed that it is necessary to introduce a relaxation factor because during the calculations the fluid domain geometry tends to converge very slowly if the total geometric deformation as a result of the structural analysis is fed in as the new geometry. In order to overcome this problem, the relaxation factor was introduced to reduce the calculated deformations to a certain amount. Thus the stable state was reached under a relaxation factor (Table 2) which was set to 0.5. It can be seen that with a relaxation factor, convergence was obtained after 6 iterations between the two codes.

Furthermore, to test both temporal and spatial resolutions, mesh and time-step refinement experiments were carried out on the first model which form case Model 1(a) and case Model 1(b). The convergence process is given in Table 3. Firstly, the mesh was nearly doubled (total node number in the fluid domain was 10374 and 19418 before and after the refinement, total node number in the arterial wall was 899 and 1715 respectively). Before the mesh refinement, the calculated maximum displacement was 0.547 mm. After the refinement, the maximum displacement was predicted to be 0.562 mm, a difference of 2.7% .

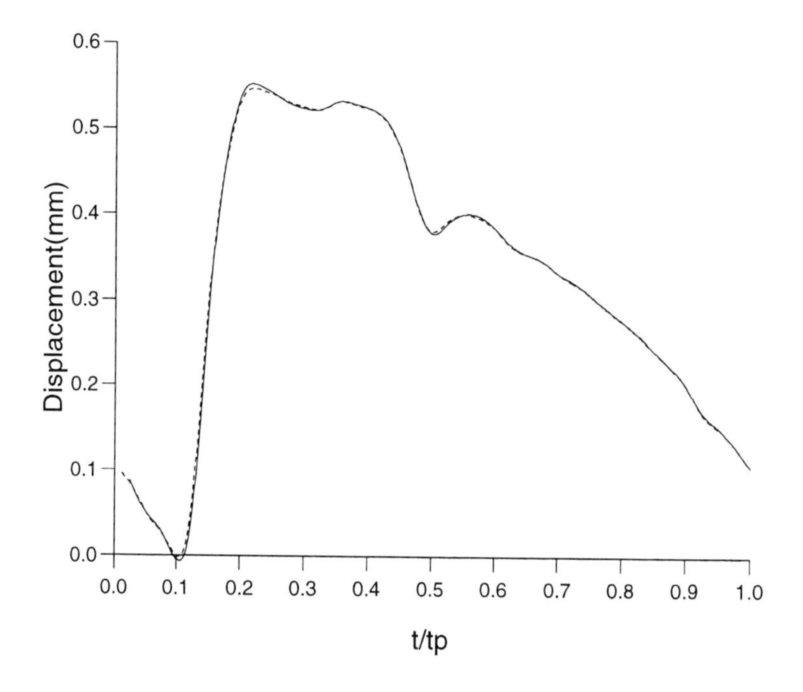

Figure 4. Comparison of displacement at the maximum displacement point of Model 1 before time step refinement without relaxation factor and after time step refinement with a relaxation factor of 0.5: solid line–before time step refinement and dashed line–after time step refinement

Secondly, the number of time steps was increased from 40 to 80 per cardiac cycle. However, for the time step refinement, convergence difficulty was again encountered. Table 4 gives the convergence process before and after relaxation. For a total of 40 time steps, the calculated maximum displacement was 0.547 mm. For 80 time steps, the maximum displacement was 0.5427 mm, showing a difference of 0.73% only.

The convergence history of the maximum displacement point before and after relaxation are plotted in Figure 3 respectively. To verify the relaxation scheme, time-dependent displacement at the maximum displacement point was compared for Model 1 and Model 1(b) with a relaxation factor as shown in Figure 4. Very good agreement has been achieved demonstrating the reliability of the relaxation technique used in this study. Since we have concentrated on the convergence problem in the paper, neither flow data nor arterial transmural stresses are given. However, the predictions provided all these information.

5 Conclusions

An advanced, three-dimensional, coupled fluid-wall model has been presented for the simulation of blood flow in compliant arterial structures. Such a coupled approach would lead to a much clearer understanding of physiological flows and vessel wall mechanics. The coupled method is capable of treating a wide range of material behaviour, such as, elastic and hyperelastic, linear and nonlinear, small and large deformations. The coupled algorithm has been applied to predictions of unsteady flow in a circular tube, with the tube material being (a) elastic, and (b) hyper-elastic [6]. Geometrically non-linear theory has been used in the displacement and stress analysis for the hyper-elastic case. The present study treats the arterial wall material as incrementally linearly elastic. The validity of this assumption in local analysis has been proven by various case studies carried out in a pressure loaded cylindrical model of a vessel having different mechanical properties.

An important issue arising from the model, that of the convergence problem, has been discussed in detail. Unfortunately, a general theory does not exist to establish the overall convergence of a coupled model. The convergence behaviour is very much problem-dependent. Numerical examples have been presented to demonstrate the validity of the proposed coupled scheme and relaxation technique.

Acknowledgement

We gratefully acknowledge the sponsorship of EPSRC.

Bibliography

1. Thubrikar, M.J., Roskelly, S.K., and Eppink, R.T. (1990). Study of stress concentration in the walls of the bovine coronary arterial branch. *J. Biomechanics*, **23**, pp. 15–26.

2. Perktold, K., and Rappitsch, G. (1994). Mathematical modeling of local arterial flow and vessel mechanics. *Computational Methods for Fluid-Structure Interaction*, Editors: J. Crolet and R. Ohayon, pp. 230–245.

3. CFX 4: User Guide (1996). UK:AEA Technology Harwell Laboratory, Oxfordshire.

4. Hibbit, Karlsson and Sorensen Inc. (1996). *ABAQUS*, **5.6**, Pawtucket, R.I. 02860.

5. Feng, Y.T., and Owen, D.R.J. (1996). Iterative solution of coupled FE/BE discretizations for plate-foundation interaction problems. *Int. J. for Numerical Methods in Engineering*, **39**, pp. 1889–1901.

6. Zhao, S.Z., Xu, X.Y., and Collins, M.W. (1997). A novel numerical method for analysis of fluid and solid coupling. *Numerical Methods in Laminar and Turbulent Flows*, **10**, Editors: C. Taylor and J. Cross, Pineridge Press, Swansea, pp. 525–534.

7. Müller, A., and Jacob, N. (1994). Explicit fluid-flow-solid interaction. *Int. J. of Computer Applications in Technology*, **7**, pp. 185–192.

8. Hofer, M., Rappitsch, G., Perktold, K., Trubel, W., and Schima, H. (1996). Numerical study of wall mechanics and fluid dynamics in end-to-side anastomoses and correlation to intimal hyperplasia. *J. Biomechanics*, **29**, pp. 1297–1308.

9. Perktold, K., and Rappitsch, G. (1995). Computer simulation of local blood flow and vessel mechanics in a compliant carotid artery bifurcation. *J. Biomechanics*, **28**, pp. 845–856.

Modelling of Cerebral Autoregulation Experiments in Humans

S.K. Kirkham*, R.E. Craine* and A.A. Birch**

**Faculty of Mathematical Studies, University of Southampton, and*
***Department of Medical Physics and Bioengineering, Southampton University Hospitals NHS Trust*

1 Introduction

Cerebral autoregulation is the process by which cerebral blood flow is kept constant despite changes in the arterial blood pressure. This process is impaired in many patient groups, however, and knowledge of a patient's ability to autoregulate could lead to an improved understanding of their condition and thereby help to manage their therapy.

The mechanisms controlling cerebral autoregulation are complex and not well understood. Current techniques for assessing it have had limited success, and development of these techniques is required in order to improve our understanding of the measurements and to establish clinical applications.

To assess autoregulation it is necessary to observe how the blood flow in the brain responds to a changing arterial blood pressure. Researchers at Southampton General Hospital (S.G.H.) have been conducting experiments using transcranial Doppler ultrasound to monitor blood velocity in the middle cerebral artery (M.C.A.), and varying the pressure by the application of lower body negative pressure. The latter is achieved at S.G.H. through the use of a vacuum box which creates an oscillating arterial pressure. The period is typically set at 12 seconds, which is, ten to twenty times longer than the period of the heart cycle. This pressure change induces an additional oscillating velocity in the middle cerebral artery. A phase lag between the pressure and velocity has been recorded and it is believed that the magnitude of this phase lag is a good indication of the status of the autoregulation processes in the brain.

The waveforms measured in the experiments are interpreted using Fourier analysis. The experimental procedure used at S.G.H. is as follows. For each vacuum box cycle, with period T_v, the zeroth and first Fourier coefficients are determined for both the pressure and maximum velocity waveforms. After k consecutive cycles of the vacuum box the Fourier coefficients are averaged, and the results are then assumed to be the waveforms produced due to the action of the vacuum box only. It is hoped that this averaging process removes any "noise" due to the underlying heart cycle with period T_h.

This paper uses a simple flow model to investigate whether this averaging process is acceptable.

2 Theoretical model

Experiments by Newell et al. [1] suggest that during large step changes in pressure the diameter of the M.C.A. does not alter and hence it seems reasonable to assume that the M.C.A. is rigid. Womersley [2] developed a mathematical model for flow in arteries based on this assumption. He introduced a number of further assumptions in order to simplify the governing equations for oscillating flow down a rigid tube. The flow is assumed to be laminar and blood is modelled as a Newtonian fluid. It is also assumed that the unsteady flow is axisymmetric and unidirectional along the axis of the tube, as is the case in simple Poisieulle flow down a tube. Introducing cylindrical polar coordinates (r, θ, z) in the usual way, with the origin situated on the pipe axis, it is assumed that the axial velocity v_z is a function of r and time, t, only.

A disadvantage of using Womersley's rigid tube model is that the pressure gradient must be prescribed. Only the pressure is measured in the experiments at S.G.H., however, and to be able to relate Womersley's rigid tube model to the experimental results it is necessary to connect the pressure gradient to the measured pressure. When autoregulation is absent it can be assumed that the pressure and pressure gradient have the same time dependence. This is possible if we assume the artery is rigid and that venous pressure is constant. Thus, when autoregulation is absent, any change in arterial pressure results in pressure changes, of slightly different magnitudes, at each end of the M.C.A, resulting in a pressure gradient which has the same time dependent form as the pressure. When autoregulation is present this assumption can no longer be made and new relationships between pressure and pressure gradient must be introduced.

As discussed in the introduction, experiments at S.G.H. involve the use of a vacuum box, which superimposes onto the heart cycle a sinusoidal pressure variation with a period of about twelve seconds. To model these experiments Womersley's model is extended in this paper to include the superposition of two pressure waveforms. Assuming that the pressure change due to the heart cycle is sinusoidal, it is assumed that the arterial pressure, p, has the form

$$p = p_0 \left(1 + \epsilon_\omega \sin(\omega t + \beta) + \epsilon_\Omega \sin(\Omega t)\right), \tag{2.1}$$

where p_0 is the average pressure in the M.C.A, $\omega = 2\pi/T_H$ is the angular frequency of the heart cycle and $\Omega = 2\pi/T_V$ is the angular frequency of the vacuum box cycle. The parameters ϵ_ω and ϵ_Ω are the magnitudes of the proportional variations in the pressure due to the heart cycle and vacuum box cycle respectively. The vacuum box is started at some random time during a heart cycle and so β reflects the amount of the heart cycle which has been completed when the vacuum box is switched on. Using this definition β satisfies $0 \leq \beta \leq 2\pi$.

In the absence of autoregulation it is assumed that the pressure gradient across the M.C.A arising from Equation (2.1) is

$$\Delta P = \Delta P_0 \left(1 + M \sin(\omega t + \beta) + N \sin(\Omega t)\right), \tag{2.2}$$

where ΔP_0 is the average pressure gradient in the M.C.A, M and N are the magnitudes of the proportional variation in the pressure gradient due to the heart and vacuum box cycles respectively. It is important to note that Womersley's model is linear so it is possible to treat the heart and vacuum box cycle terms separately and then add them together to obtain the final solution. Using Equation (2.2) in Womersley's model it is easy to show that the axial velocity has the form

$$v_z = v_0 \left((1 - R^2) + \Gamma_\omega M \sin(\omega t - \phi_\omega + \beta) + \Gamma_\Omega N \sin(\Omega t - \phi_\Omega)\right). \tag{2.3}$$

The underlying steady flow component is given by $v_0 = \Delta P_0 a^2 / 4\mu$ where a is the radius of the M.C.A. and μ is the viscosity of blood. In Equation (2.3) the variable $R = r/a$ is the non-dimensional radial coordinate, Γ_ω is a factor relating the proportional change in the amplitude of the flow velocity to the corresponding change in the amplitude of the pressure gradient and ϕ_ω is the phase lag between the pressure gradient and velocity. Both Γ_ω and ϕ_ω linked to the heart cycle depend on R and the non-dimensional Womersley parameter, $\alpha = a\sqrt{\omega/\nu}$, where ν is the kinematic viscosity. The quantities Γ_Ω and ϕ_Ω denote the corresponding parameters for the vacuum box cycle. The experiments at S.G.H. measure the maximum velocity, which occurs in the centre of the tube where $R = 0$.

3 Fourier decomposition of pressure and maximum velocity waveforms

In order to assess the influence of the underlying heart cycle on the waveforms associated with the vacuum box cycle, it is necessary to express the heart cycle terms in Equations (2.1) and (2.3) as Fourier series over the period of the vacuum box. Including only the first two terms in these series, Equations (2.1) and (2.3) can be written as

$$p = p_o \left(1 + \frac{a_o}{2} + N_p \sin(\Omega t - \phi_p)\right), \tag{3.1}$$

$$v_{max} = v_o \left(1 + \frac{a_{0v}}{2} + N_v \sin(\Omega t - \phi_v)\right), \tag{3.2}$$

where

$$
\begin{aligned}
v_{max} &= v_z(R=0),\\
N_p &= (\epsilon_\Omega^2 + a_1^2 + b_1^2 + 2b_1\epsilon_\Omega)^{\frac{1}{2}},\\
\phi_p &= \tan^{-1}\left(\frac{-a_1}{\epsilon_\Omega + b_1}\right),\\
N_v &= \left(a_{1v}^2 + b_{1v}^2 + \Gamma_\Omega^2 N^2 - 2\Gamma_\Omega N\{a_{1v}\sin\phi_\Omega - b_{1v}\cos\phi_\Omega\}\right)^{\frac{1}{2}},\\
\phi_v &= \tan^{-1}\left(\frac{-a_{1v} + \Gamma_\Omega N\sin\phi_\Omega}{b_{1v} + \Gamma_\Omega N\cos\phi_\Omega}\right).
\end{aligned}
$$

Note that with the usual notation, a_0, a_1 and b_1 are the leading Fourier coefficients of $\epsilon_\omega \sin(\omega t+\beta)$ and a_{0v}, a_{1v} and b_{1v} are the corresponding coefficients of $\Gamma_\omega M\sin(\omega t + \beta - \phi_\omega)$. Higher Fourier coefficients are ignored since they are not currently used in the analysis of the experimental data. Observe from Equations (3.1) and (3.2) that N_p and N_v are proportional to the magnitudes of the oscillations of the pressure and maximum velocity, respectively. When the heart cycle is ignored (i.e. $a_0 = a_1 = b_1 = a_{0v} = a_{1v} = b_{1v} = 0$) it is easily seen that the general expressions reduce to $N_p = \epsilon_\Omega$ and $N_v = \Gamma_\Omega N$. The angle ϕ_p represents the phase lag between the pressure waveforms when the heart cycle is included and when it is ignored, and it follows that $\phi_p = 0$ when the heart cycle is ignored. The angle ϕ_v represents the corresponding phase lag for the maximum velocity waveforms and clearly equals ϕ_Ω when the heart cycle is ignored. The phase lag measured in the experiments is $\phi_p - \phi_v$.

For the results presented in this section we introduce numerical values for the physical parameters that are appropriate for blood flow in the M.C.A. It is assumed that $a = 1.5$ mm and $\nu = 3.8 \times 10^{-6}\mathrm{m}^2\mathrm{s}^{-1}$. Typical values for ϵ_ω and ϵ_Ω are obtained from the experimental measurements of the pressure taken at S.G.H. and are 0.3 and 0.05, respectively. The period of the vacuum box is assumed to be 12 seconds (i.e. $\Omega = \pi/6$) and the values $M = 0.35$ and $N = 0.05$ are prescribed so that the magnitudes of the resulting variations in velocity are similar to those observed in the experiments.

Figures 1 and 2 show the variation of N_v and ϕ_v, with $p = \omega/\Omega$ for three values of β. Note that if the vacuum box was acting alone then N_v and ϕ_v would have the constant values 0.05 and 0.055 respectively. Figure 1 shows that the heart cycle can affect N_v by as much as 6%. Figure 2 reveals that ϕ_v also shows a significant variation due to the heart cycle. It can take a value in the range -0.28 to 0.38 radians, or -16° to 22°, depending on the values of p and the initial phase difference β.

Figure 3 shows the variation of $\phi_p - \phi_v$ which is the phase lag measured in experiments. The value of this phase lag is also affected by the heart cycle and it can take a value in the range -0.18 to 0.15 radians, or -10° to 8°, depending on the values of p and β.

The results presented in Figures 1–3 clearly reveal that if only one period of the vacuum box cycle is considered then the underlying heart cycle has a

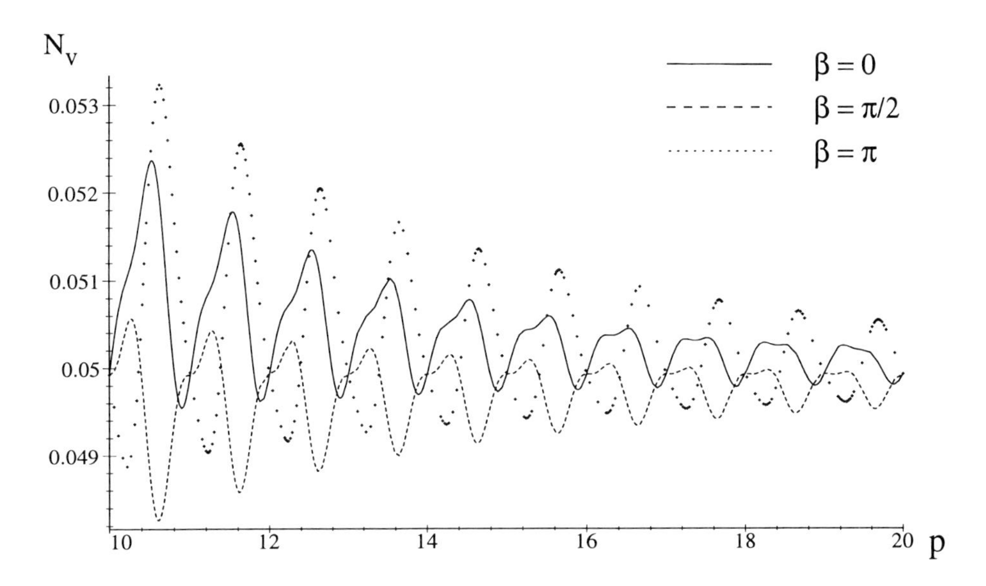

Figure 1. Variation of N_v, the magnitude of oscillation of the maximum velocity, with p (where $\omega = p\Omega = \pi p/6$) for three values of β

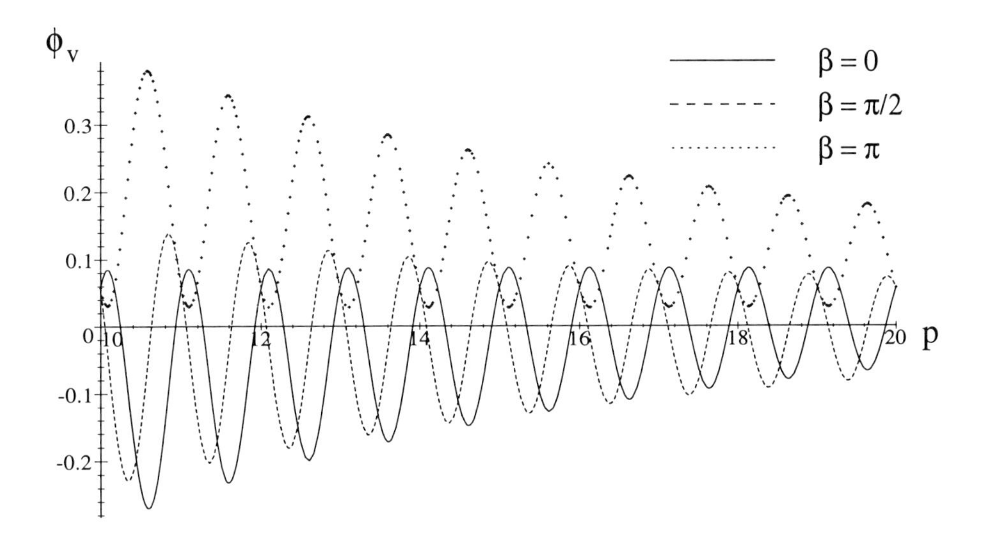

Figure 2. Variation of ϕ_v, the phase shift in the maximum velocity, with p (where $\omega = p\Omega = \pi p/6$) for three values of β

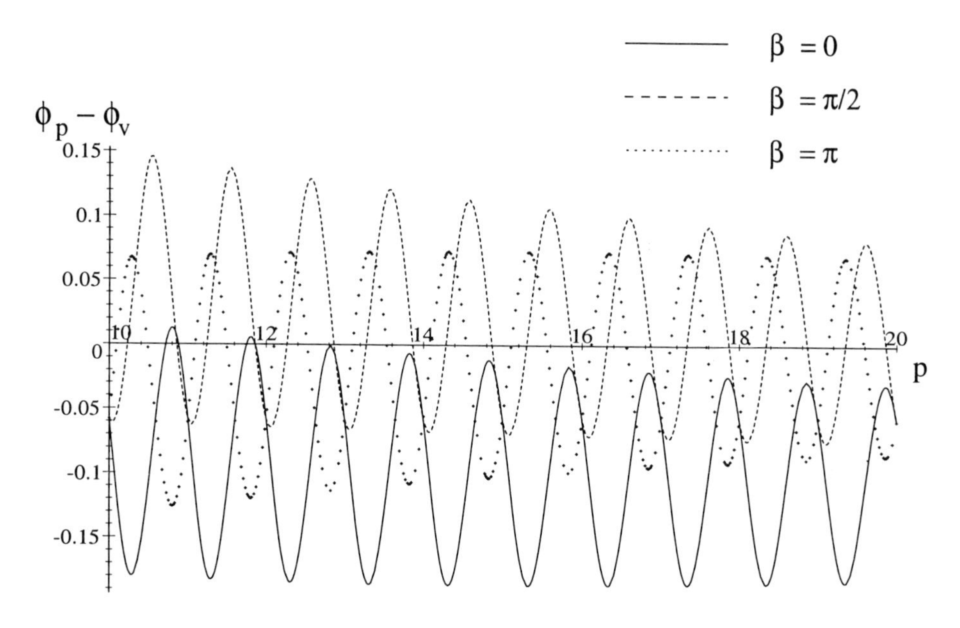

Figure 3. Variation of the $\phi_p - \phi_v$, the phase lag, with p (where $\omega = p\Omega = \pi p/6$) for three values of β

significant effect on the Fourier decomposition of the velocity over the vacuum box cycle. It should be noted that, in a similar way, the heart cycle affects the magnitude and phase of the pressure, but the results are omitted here.

In order to consider the effect of the averaging process we consider k consecutive vacuum box cycles and set $p = 12 + r$ where $0 \leq r \leq 1$. This is equivalent to considering a vacuum box with a twelve second period and a heart cycle period between 0.92 and 1 second.

When the value of β is fixed at the beginning of the first cycle of the vacuum box and the values of ω and Ω are known, it is easy to see that the values of β for the second and subsequent cycles of the vacuum box are prescribed. On using this observation, the results obtained with our model can be extended in a straightforward way to k consecutive cycles of the vacuum box. The values of N_v and ϕ_v can be calculated for each vacuum box cycle and the average taken. The final solutions retain the forms given in Equations (3.1) and (3.2) but the new averaged values of the parameters are denoted by a superposed bar.

Figures 4 and 5 show the variation of $\overline{N}_v$ and $\overline{\phi}_v$ with r for the cases when $k = 10$ and $k = 25$, for the particular choice $p = 12 + r$ and $\beta = 0$. Both figures also show the corresponding variations of N_v and ϕ_v obtained for a single vacuum box cycle, which are equivalent to the graphs shown in Figures 1 and 2 when $\beta = 0$ and $12 \leq p \leq 13$. It is clear from Figures 4 and 5 that both $\overline{N}_v$

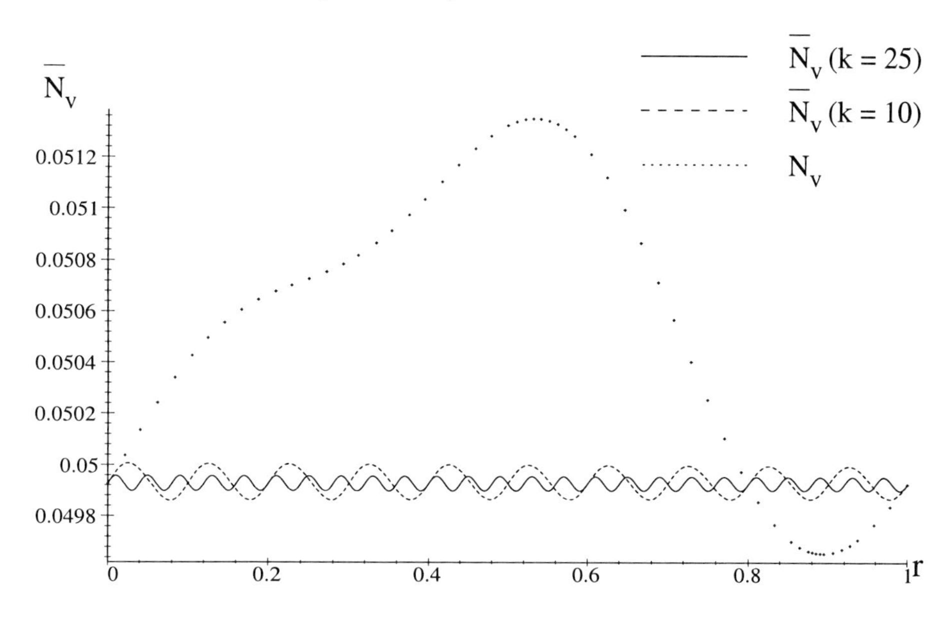

Figure 4. Variation of $\overline{N}_v$, the averaged magnitude of the maximum velocity, with r (when $\beta = 0$ and $\omega = (12 + r)\Omega$)

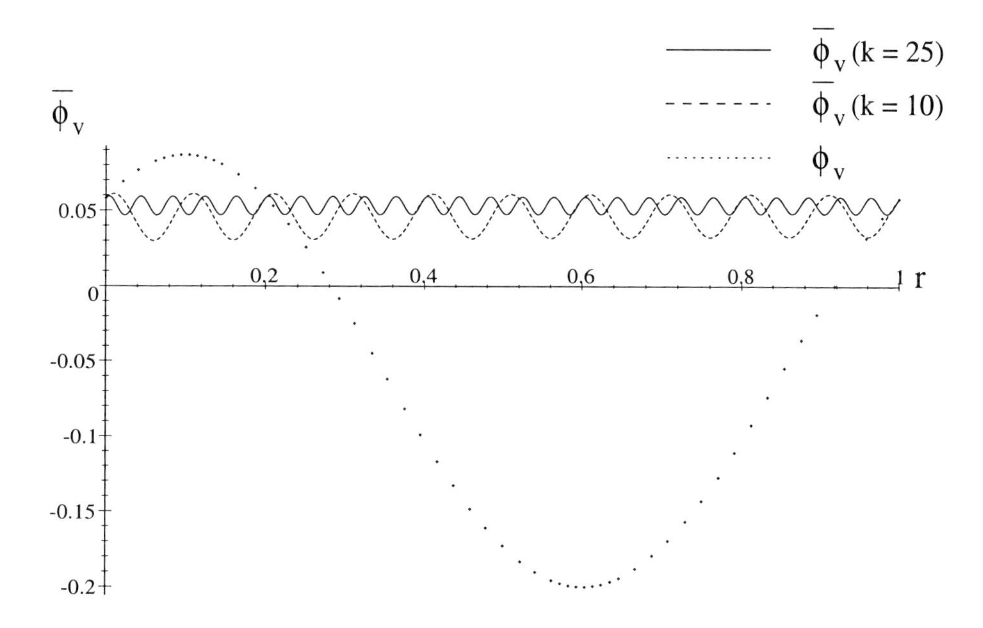

Figure 5. Variation of $\overline{\phi}_v$, the averaged phase lag of the maximum velocity phase lag, with r (when $\beta = 0$ and $\omega = (12 + r)\Omega$)

and $\overline{\phi}_v$ show considerably less variation from the constant values $\Gamma_\Omega N$ and ϕ_Ω (obtained when the heart cycle is ignored) than either N_v or ϕ_v.

It is obvious from Figures 4 and 5 that averaging over 25 vacuum box cycles reduces the influence of the underlying heart cycle on the variation in magnitude of the maximum velocity from as much as 6% to less than 1%. The influence of the heart cycle on the phase lag is similarly reduced, with changes of up to 18° falling to less than 1°.

It is important to emphasize that a large number of other results were calculated with different values of Ω, p and β. The conclusions drawn from all these results agree qualitatively with those discussed above.

4 Concluding remarks

The accuracy of the Fourier analysis procedure used in assessing the data obtained in the experiments on the M.C.A. at S.G.H. has been investigated in this paper by using an extension of Womersley's rigid tube model. Our results suggest that in the absence of autoregulation, the neglect of the heart cycle could lead to significant differences in the Fourier coefficients if only one cycle of the vacuum was considered. However, these errors have been shown to become very small when averaging over 25 consecutive cycles of the vacuum box, and so the procedure used in the experiments at S.G.H. seems appropriate.

The assumptions introduced in our mathematical model are thought to be reasonable so it is believed that our general conclusions would remain valid if more accurate arterial models were used.

Finally it should be emphasized that, for simplicity, autoregulation was omitted from our model. Including the latter complicates the mathematics but it is expected that the averaging procedure would again produce results which are little affected by the underlying heart cycle.

Bibliography

1. Newell, D.W., Aaslid, R., Lam, A., Mayberg, T.S., and Winn, H.R. (1994). *Stroke*, **25**, p. 340.

2. Womersley, J.R. (1955). *Phil. Mag.*, **46**, p. 199.

The Numerical Investigation of a Novel Haemodynamic Control Device which Reduces the Development of Occlusive Arterial Intimal Hyperplasia

J.S. Cole*, M.A. Gillan*, S. Raghunathan*, A. McKinley and M.J.G. O'Reilly****

**School of Aeronautical Engineering, Queen's University of Belfast, Northern Ireland, and ** Vascular Unit, Belfast City Hospital, Northern Ireland*

Abstract

A computational fluid dynamics investigation of the pulsatile, non-Newtonian blood flow through a typical, three-dimensional model of a human femorodistal bypass graft has emphasised the complicated temporal and spatial dependence of the flow patterns at the proximal and distal anastomoses.

A novel, simply configured Haemodynamic Control Device (HCD), fitted at the ends of the graft, judiciously modified the anastomotic flow fields. Flow separation and recirculation were diminished. It is believed that the successful optimisation of the HCD would have important benefits for vascular bypass patients since the inhibiting of the unnatural, disturbed haemodynamics should mean that the development of intimal hyperplasia, and eventual graft failure, is less likely.

1 Introduction

The bypassing of critically stenosed arteries using biological or synthetic grafts is a common surgical procedure for alleviating life-threatening angina or limb-threatening ischaemia. However, intimal hyperplasia, the progressive, abnormal proliferation of the innermost layer of the arterial wall, is a major cause of the early and medium term failure of bypass grafts [1]. It is known that intimal hyperplasia develops exclusively at the suture line where the graft is surgically attached to the host artery, and on the floor of the host artery opposite the anastomosis [2]. The disease occurs predominantly at the distal rather than the proximal anastomosis, with prosthetic grafts being at greater risk than biological grafts. Figure 1 outlines a typical vascular bypass configuration with the observed distribution of anastomotic intimal hyperplasia.

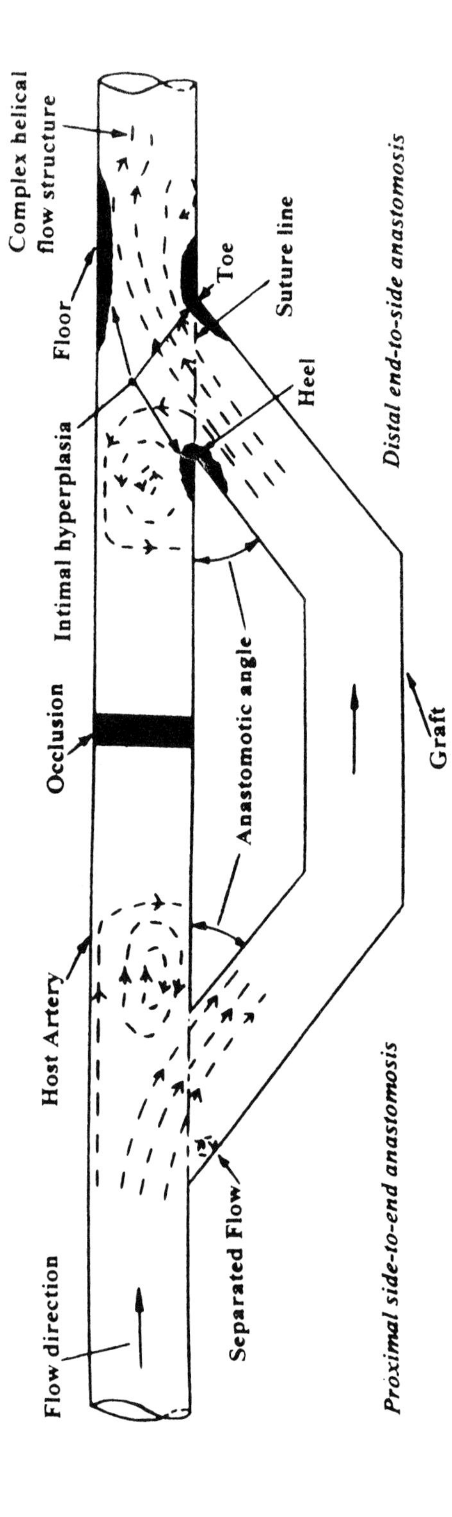

Figure 1. Schematic haemodynamic flow pattern for a bypass graft indicating regions of intimal hyperplasia in the vicinity of the distal end-to-side anastomosis

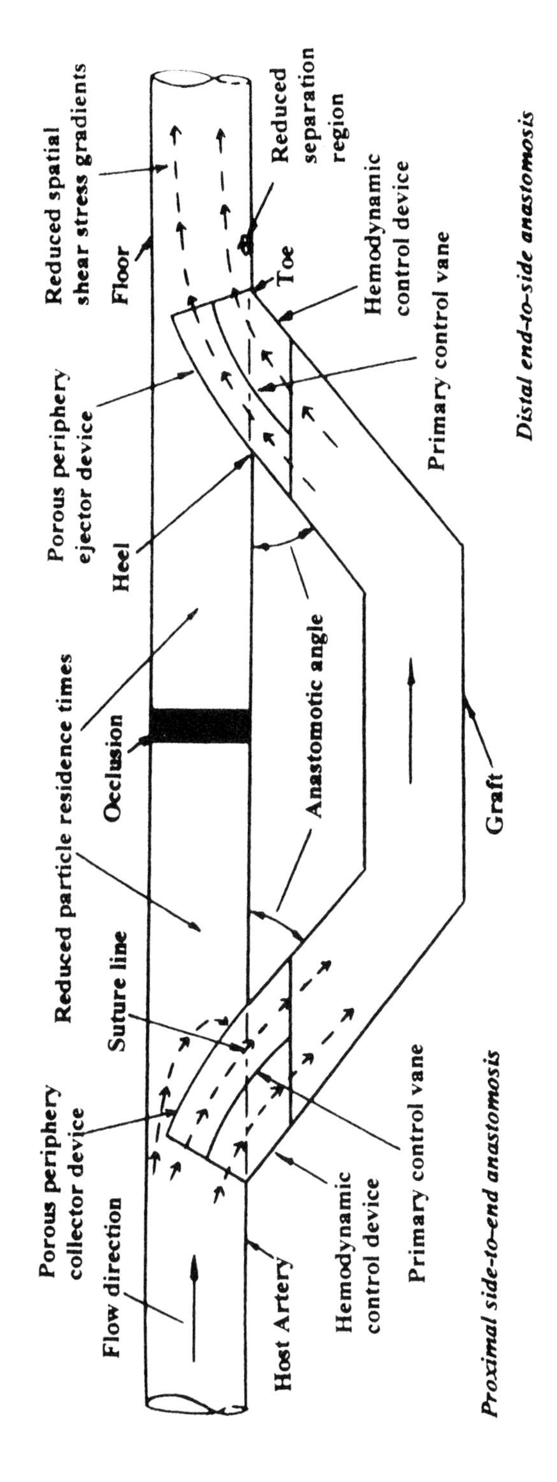

Figure 2. Schematic haemodynamic flow pattern for a bypass graft with a novel haemodynamic control device (HCD) fitted at both anastomoses

While the exact pathogenic mechanisms for intimal hyperplasia are unclear, it is thought that a number of interrelating factors are involved. These include the compliance mismatch between the graft and artery [3], endothelial injury, and the interaction between blood particles and the arterial wall [1, 4]. Furthermore, given the focal nature of anastomotic intimal hyperplasia, it is believed that haemodynamic influences are important [5].

Much research has focused on optimising the graft material [6], analysing anastomotic flow patterns [7, 8], or incorporating vein patches into the anastomoses [9]. This study considers the possibility of haemodynamic control at the anastomoses. It is proposed that the attachment of a Haemodynamic Control Device (HCD) [10] at the anastomoses can adapt the local flow fields, ensuring improved flow patterns, and thereby reducing the likelihood of the progression of intimal hyperplasia and subsequent graft failure. Figure 2 displays a typical bypass configuration with HCDs fitted at both the proximal and distal anastomoses.

The biocompatible, synthetic HCD is of one-piece construction and can accommodate a primary control vane, optional secondary control vanes and an optional periphery collector/ejector device. The primary control vane is a thin surface used to enhance flow vectoring into and out of the graft. Secondary control vanes may be utilised in larger diameter grafts. The periphery collector/ejector device is another thin haemodynamic flow control surface, whose porosity may be optimised to secure reduced particle residence times in the locality of the occlusion. The size, profile and pitch of the components of the HCD are chosen in order to conform to host artery elasticity and graft/host artery architecture. It is expected that the HCDs would be attached to the graft during the surgical procedure using either suturing techniques or biological glues.

In this investigation, numerical simulations of the three-dimensional, pulsatile, non-Newtonian flow of blood through a typical human femorodistal bypass have been performed. In contrast to previous studies, flow through the complete bypass model has been computed, rather than that at either the proximal or distal anastomosis in isolation. The ability of a simply configured HCD to modify the anastomotic flow fields is examined.

2 Model geometry

The symmetry plane of the standard, rigid bypass model is depicted in Figure 3. The host artery has an internal diameter of 8 mm and is representative of a fully occluded, human femoral artery. The bypass graft is a circular tube of internal diameter 6 mm, symbolising a synthetic graft. It is attached at an angle of 30° to the host artery at both anastomoses. The graft is of length 16 cm and it possesses smooth curvature in the symmetry plane.

It is claimed [11] that the idealisation of rigid arterial walls can be made when only local flow patterns in short segments of large arteries are of interest. These conditions prevail in the current study.

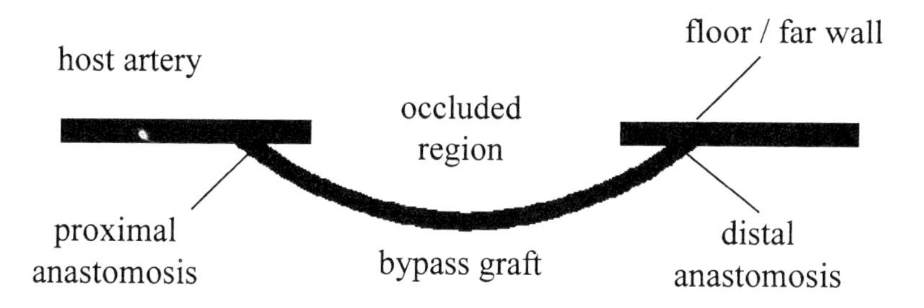

Figure 3. Symmetry plane of the bypass model

For the bypass model including the attached HCDs, the host artery and graft geometries were identical to those of the standard configuration. Each HCD is constructed as a single, thin, rigid, solid surface (the collector/ejector device), extending smoothly from the inner graft wall into the lumen of the host artery. The HCD is sized so that under the conditions of typical viscoelastic arterial wall motion, there would not be any interference between the device and the wall.

3 Numerical model and flow conditions

Blood is assumed to be an incompressible, non-Newtonian fluid of density 1050 kgm^{-3} and viscosity, μ_{eff}, given by the power law

$$\mu_{eff} = k\dot{\gamma}^{n-1}$$

where $k = 0.042$ Pas^n and $n = 0.61$. Blood flow through the femorodistal bypass possesses the attributes of three-dimensional, time-dependent, incompressible, isothermal, laminar flow. The governing equations for such a flow are

$$\begin{aligned} \nabla.\mathbf{u} &= 0 && \text{continuity equation} \\ \rho\frac{\partial \mathbf{u}}{\partial t} + \rho(\mathbf{u}.\nabla)\mathbf{u} &= -\nabla p + \nabla.(\mu_{eff}\nabla\mathbf{u}) && \text{Navier-Stokes equations.} \end{aligned}$$

Numerical computations were performed using the commercial finite volume Reynolds-Averaged Navier-Stokes code, FLUENT [12]. The fully structured physical mesh overlaying the standard bypass model is displayed in Figure 4(a). Figure 4(b) shows an enlarged view of the proximal junction with the HCD fixed at the end of the graft. Simulations were conducted for the bypass models with and without HCDs in place, under identical flow conditions.

At the inlet boundary, a time-dependent inflow velocity, calculated from the representative femoral artery flow wave [13] (Figure 5), was specified. The mean flow rate of 2.25 mls^{-1} corresponds to a mean inflow velocity of 4.48 cms^{-1} or a mean inflow Reynolds number (based on host artery diameter and a reference

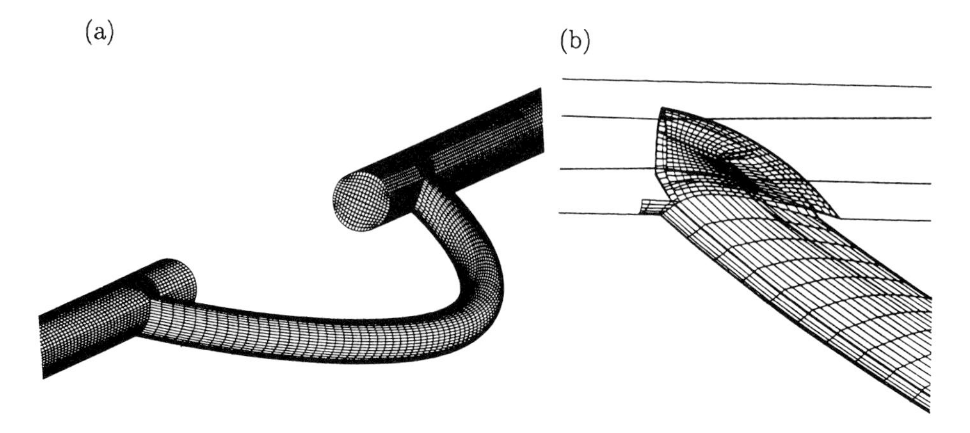

Figure 4. (a) Physical grid overlaying the bypass model, and (b) Haemodynamic control device at the proximal anastomosis

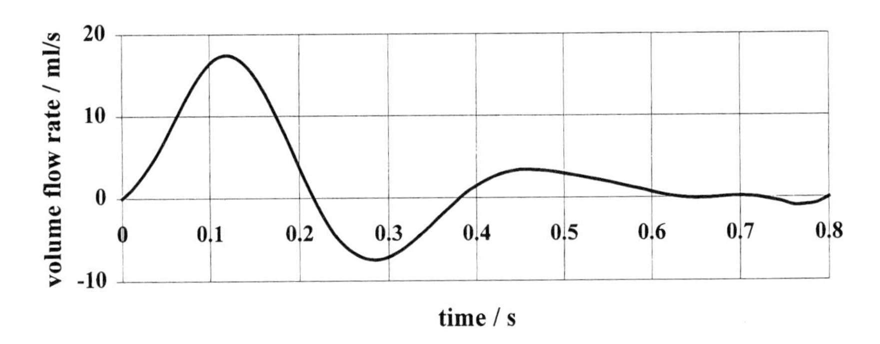

Figure 5. Femoral artery flow wave

blood viscosity of 0.0035 $\mathrm{kgm}^{-1}s^{-1}$) of 107. The maximum Reynolds number during the cycle is 830. The length of the inflow segment of the host artery is such that fully developed boundary layer profiles exist throughout the cycle on approach to the proximal anastomosis. FLUENT's OUTLET condition was applied at the outflow boundary, enforcing fully developed flow at that location.

The numerical model had previously been validated by computing steady and pulsatile flows of blood through a generic carotid artery bifurcation model [14], excellent agreement being attained with published results [15].

4 Results and discussion

The simulation has confirmed that the flow of blood through the standard bypass model is extremely complicated, exhibiting significant temporal and spatial dependence. Figure 6 illustrates the flow field in the bypass symmetry plane at various stages during the cardiac cycle, and provides a comparison with the corresponding flow patterns in the model with HCDs fitted.

At the maximum flow rate in the standard model (Figures 6(a) and 6(b)), the blood separates from the floor of the host artery on approach to the proximal anastomosis. A fraction of the flow travels past the junction in the direction of the occlusion, before following curved paths, moving above and below the symmetry plane, and entering the graft. On its passage through the bypass, the blood experiences centrifugal forces, so that the velocity profiles are skewed towards the outer wall as the fluid nears the distal anastomosis. On leaving the graft, blood particles are forced across the lumen, separation occurring at the toe. A large zone of low momentum, recirculating fluid is observed between the junction and the occlusion. Significant secondary motion is detected within the distal host artery as the blood flows downstream along helical paths.

The corresponding plots obtained under controlled conditions reveal the striking influences of the HCDs on the flow field (Figures 6(a) and 6(b)). At the proximal anastomosis, the fluid is ducted smoothly into the graft, separation being avoided. The separation region at the distal anastomosis is reduced in size, and commences further downstream from the toe. Recirculation along the floor of the distal host artery is greatly decreased.

During the reversed flow phase of the cycle (Figure 6(c)), fluid particles follow convoluted paths at the distal anastomosis in the standard model as the vortex, which existed next to the occlusion, has relocated at the graft/artery interface. However, the presence of the HCD ensures reduced flow disturbance as the development of a vortex across the anastomosis is prevented.

The effects of the HCD during diastole are also clear (Figure 6(d)). Recirculation opposite the distal anastomosis is reduced and the flow pathlines are smoother, interaction between the mainstream and fluid in the segment between the junction and the occlusion having been inhibited.

Secondary motion within the bypass and the distal host artery varied in strength during the cycle, being most notable during systolic deceleration. Different secondary flow patterns were observed in the standard and modified models due to the constraint imposed on the fluid exiting the graft in the latter case.

Proposed mechanisms for intimal hyperplasia often implicate an adverse shear stress distribution on the arterial floor opposite the distal anastomosis. This analysis revealed large shear stresses and unnaturally large spatial gradients of shear stress on the floor of the distal host artery during systole in the standard model. When the HCD was in place, corresponding shear stress magnitudes were reduced, while the desired effect of decreased spatial shear stress gradients was also achieved during the greater part of the pulse cycle.

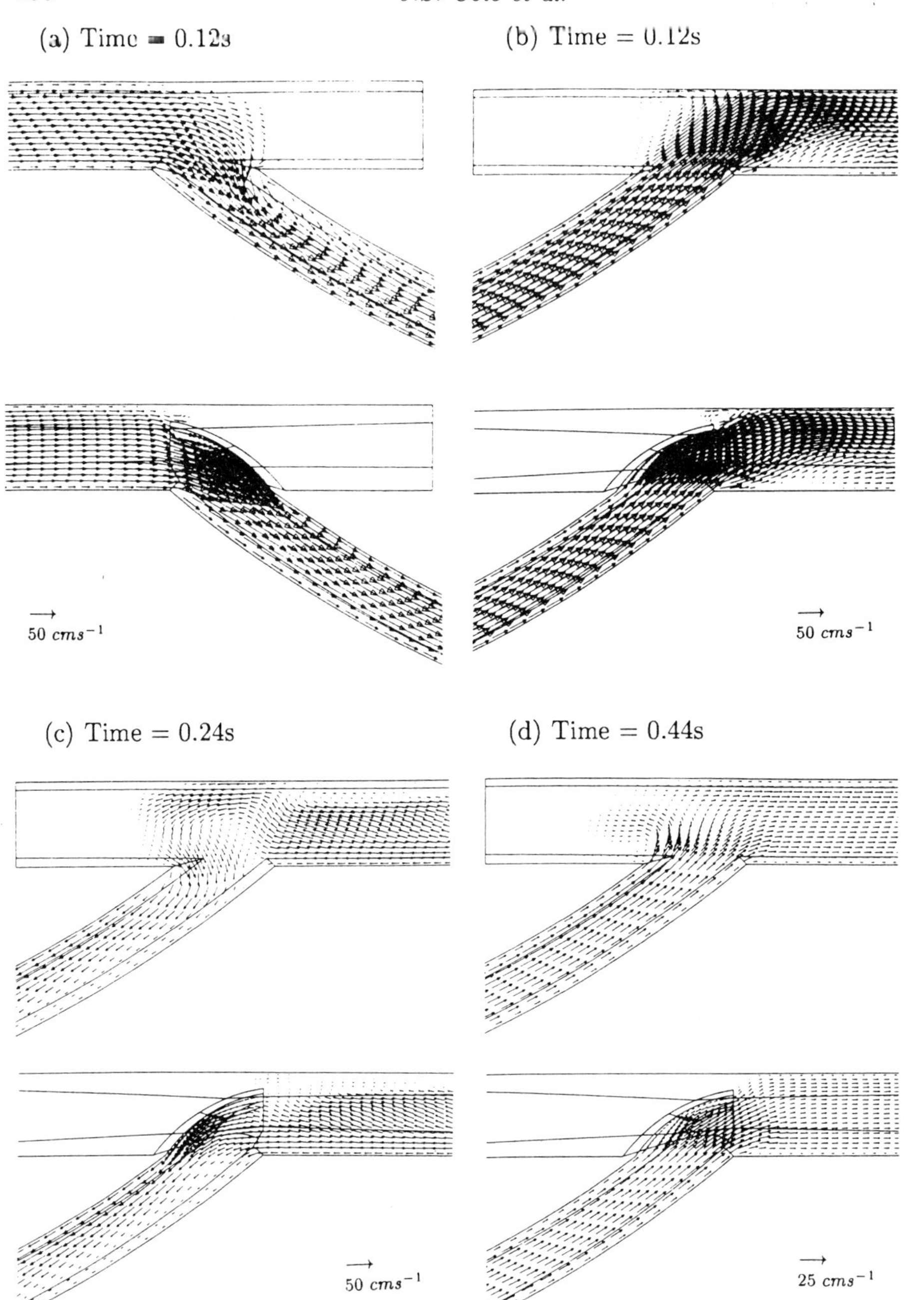

Figure 6. Comparison of the flow fields in the symmetry planes of the standard and modified bypass models at various times during the pulse cycle. (a) and (b) Maximum flow rate, (c) Reversed flow phase, and (d) Diastole

5 Conclusions

The computation of the flow of blood through a typical femorodistal bypass model has highlighted the extremely complex anastomotic flow patterns, and their significant temporal and spatial variations. The undesirable features of flow separation, recirculation, and convoluted pathlines, and the consequential elevated particle residence times, are especially evident at the distal anastomosis. The importance of modelling the complete bypass system was confirmed, since the flow at the distal junction was shown to be influenced by that upstream.

The HCDs at the anastomoses judiciously adapted the flow field. It is concluded that a successful optimisation of the HCD could have far-reaching benefits in vascular surgery. The streamlining of the flow, and the restricting of disturbed haemodynamics, should diminish the likelihood of the progression of intimal hyperplasia and eventual graft failure.

Bibliography

1. Chervu, A., and Moore, W. (1990). An overview of intimal hyperplasia. *Surg., Gyn. and Obstet.*, **171**, pp. 433–447.

2. Sottiurai, V.S., Yao, J.S.T., Batson, R.C., Lim Sue, S., Jones, R., and Nakamura, Y.A. (1989). Distal anastomotic intimal hyperplasia: histopathologic character and biogenesis. *Ann. Vasc. Surg.*, **3**, pp. 26–33.

3. Abbott, W.M., Megerman, J., Hasson, J.E., L'Italien, G., and Warnock, D.F. (1987). Effect of compliance mismatch on vascular graft patency. *J. Vasc. Surg.*, **5**, pp. 376–382.

4. Golledge, J. (1997). Vein grafts: haemodynamic forces on the endothelium - a review. *Eur. J. Vasc. Endovasc. Surg.*, **14**, pp. 333–343.

5. Bassiouny, H.S., White, S., Glagov, S., Choi, E., Giddens, D.P., and Zarins, C.K. (1992). Anastomotic intimal hyperplasia: mechanical injury or flow induced? *J. Vasc. Surg.*, **15**, pp. 708–717.

6. Greisler, H.P. (1991). *New Biologic and Synthetic Vascular Prostheses* (1st Edition). R.G. Landes Company, Austin, Texas.

7. Hughes, P.E., and How, T.V. (1996). Effects of geometry and flow division on flow structures in models of the distal end-to-side anastomosis. *J. Biomech.*, **29**, pp. 855–872.

8. Hofer, M., Rappitsch, G., Perktold, K., Trubel, W., and Schima, H. (1996). Numerical study of wall mechanics and fluid dynamics in end-to-side anastomoses and correlation to intimal hyperplasia. *J. Biomech.*, **29**, pp. 1297–1308.

9. Taylor, R.S., Loh, A., McFarland, R.J., Cox, M., and Chester, J.F. (1992). Improved technique for polytetrafluoroethylene bypass grafting: long-term results using anastomotic vein patches. *Br. J. Surg.*, **79**, pp. 348–354.

10. Gillan, M.A., and The Queen's University of Belfast. (1997). Haemodynamic control device. *British Patent Application No. 9706965.2.*

11. Zarins, C.K., Giddens, D.P., Bharadvaj, B.K., Sottiurai, V.S., Mabon, R.F., and Glagov, S. (1983). Carotid bifurcation atherosclerosis - quantitative correlation of plaque localization with flow velocity profiles and wall shear stress. *Circ. Res.*, **53**, pp. 502–514.

12. (1996). *FLUENT User's Guide (Release 4.4.).* Fluent Incorporated, Lebanon, New Hampshire.

13. Steinman, D.A., Vinh, B., Ethier, C.R., Ojha, M., Cobbold, R.S.C., and Johnston, K.W. (1993). A numerical simulation of flow in a two-dimensional end-to-side anastomosis model. *ASME. J. Biomech. Eng.*, **115**, pp. 112–118.

14. Cole, J.S., Gillan, M.A., and Raghunathan, S. (1998). A CFD study of steady and pulsatile flows within an arterial bifurcation. *36th AIAA Aerospace Sciences Meeting and Exhibit. AIAA Paper 98-0790.*

15. Ku, D.N., and Giddens, D.P. (1987). Laser-Doppler anemometer measurements of pulsatile flow in a model carotid bifurcation. *J. Biomech.*, **20**, pp. 407–421.

Hemorheology and Hemodynamics: A Complex Approach to the Coronary Circulation and Development of Myocardial Ischemia

K. Toth*, **G. Kesmarky***, **L. Habon****, **G. Vajda****, **T. Habon*** and **I. Juricskay***

**Department of Medicine, Division of Cardiology, and **Department of Surgery, Heart Center, University Medical School of Pecs, Hungary*

Abstract

In the developed countries cardiovascular diseases, and among these ischemic heart disease (IHD), are the most frequent causes of mortality. While the classical risk factors of IHD (age, male sex, smoking, high cholesterol level, hypertension, diabetes, stress, etc.) are well known for decades, the important role of hemorheological parameters became really evident only during the last two decades when researchers verified that hemodynamic and hemorheologic factors play an almost equally important role in the coronary circulation. It is still questioned if these parameters have primary or secondary significance in the development of these diseases, but the results of recent studies show that these factors are primary risk factors in ischemic heart disease. However, it is still not completely clear, if these parameters change parallel with the severity of the disease during the clinical course, if they can be positively modified as a result of the adequate treatment, and how they affect the prognosis. Our own data on close to 1000 patients with IHD and acute myocardial infarction (AMI) show that the rheological parameters in IHD and AMI are in the pathological range and are significantly higher than in healthy controls.

In one of our latest studies, the relation of these parameters to the severity of coronary artery disease (CAD) was examined. The data of 162 patients (mean age: 55 ± 10 years) undergoing coronary angiography and 59 healthy controls (mean age: 35 ± 10 years) were analyzed. Hemorheological parameters (hematocrit, fibrinogen level, plasma viscosity (PV) and apparent whole blood viscosity (WBV)) were determined and circulatory index (CRI) was calculated. Patients were classified into three groups according to their coronary vessel state based on the coronary angiogram: Group 1 ($n = 31$, mean age: 51 ± 10 years) without significant CAD, Group 2 ($n = 29$, mean age: 51 ± 10 years) with single vessel disease, Group 3 ($n = 102$, mean age: 57 ± 9 years) with multivessel disease. All the measured hemorheological parameters of IHD patients were significantly higher than those of controls ($p < 0.05$-0.001). Fibrinogen and PV was significantly elevated in Groups 2 and 3 comparing to Group 1 ($p < 0.05$ and 0.001). Hematocrit and WBV was significantly increased in Group 3 comparing to Groups 1 and 2 ($p < 0.02$ and 0.01). CRI was significantly decreased in IHD patients, and it was also lower in Group 3 than in Group 2 ($p < 0.05$).

All these findings indicate that hemorheological parameters may play a role in the pathogenesis and development of CAD.

1 Introduction

Ischemic heart disease (IHD) is responsible for the greatest part of total morbidity and mortality in all the developed countries. The pathophysiological basis of the development of myocardial ischemia is an imbalance between the oxygen supply and oxygen demand of the myocardium. In most cases the diminished oxygen supply results from impaired perfusion in the coronary arteries as a result of coronary artery disease (CAD), the stenotic lesion of one or more coronary branches.

In the regulation of coronary circulation both hemodynamic and hemorheologic factors play an important role (Figure 1). The central hemodynamic parameters of patients with ischemic heart disease at rest generally do not differ significantly from normals in the earlier stages of IHD, but the hemorheological factors of these patients can be altered, and significantly worse than those of healthy persons. In several studies hemorheological parameters were proved to be primary cardiovascular risk factors [1–6,8,9,11–21,25,26]. The coronary vessel system is a special part of the circulation, since there is a continuous change in blood flow, perfusion pressure and shear rate due to the cardiac cycle. Therefore the role of rheological alterations may be of higher importance than in other parts of the circulatory system. In previous studies on coronary artery disease data were not unequivocal, whether any of the hemorheological factors are in correlation with the stages of CAD, and if so, then which of these parameters shows the correlation [11–17].

In our previous studies we evaluated the hemodynamic and hemorheologic parameters of patients with IHD and healthy controls, and introduced a new index - Circulatory Index <CRI> - for the characterization of the hemodynamic and hemorheologic conditions. During these studies the data of more than 500

Vascular component

Myocardial component

CORONARY BLOOD FLOW

Perfusion pressure

Blood viscosity

Figure 1. The regulation of coronary blood flow

patients with ischemic heart disease were compared to close to 100 healthy controls. These studies verified that all the measured hemorheological parameters (hematocrit, white blood cell count, fibrinogen level, plasma and whole blood viscosity) were significantly higher in patients with IHD than in healthy controls, while central hemodynamic parameters at rest did not differ significantly from normals. The newly introduced parameter - circulatory index - showed also a highly significant difference between IHD patients and controls, and was more sensitive than hemorheological or hemodynamical parameters alone [21,22].

In one of our ongoing studies the relation between hemorheological parameters and severity of CAD is examined. The hemodynamic state is also measured and the circulatory index calculated.

2 Patients and methods

Up to now, 162 IHD patients (mean age: 55 ± 10 years) and 59 healthy controls (Group 0 (*G*-0), mean age: 35 ± 10 years) were examined in this study. All the patients had angina pectoris in their clinical history. After the non-invasive diagnostic procedures (ECG, echocardiography, stress tests, myocardial perfusion scintigraphy) which proved the IHD, all patients underwent coronary angiography. They were classified into three groups according to their coronary vessel state based on the coronary angiogram. The first group (*G*-1) included 31 patients (mean age: 51 ± 10 years) without significant coronary artery disease (less than 70% stenosis) in spite of the positive non-invasive tests. Patients with definite coronary artery spasm or coronary X syndrome (small vessel disease) also belonged to this group. In the second group (*G*-2) there were 29 patients (mean age: 51 ± 10 years) with significant lesion (stenosis or occlusion) of one vessel. In the third group (*G*-3) there were 102 patients (mean age: 57 ± 9 yrs) with severe multiple stenoses or occlusions of more than one coronary arteries (double and triple vessel disease). All of our patients were on combined antianginal-antiischemic drug therapy, including nitrates, beta-blockers, calcium-antagonists, ASA, ACE-inhibitors, lipid lowering agents.

Blood samples were taken from the cubital vein, and routine blood chemistry and hemorheological parameters - hematocrit, plasma fibrinogen level, plasma and whole blood viscosity - were determined. Hematocrit was measured with microhematocrit centrifuge. Whole blood and plasma viscosities were determined on Hevimet 40 capillary viscosimeter (Hemorex, Hungary). In this viscosimeter the flow of whole blood (or plasma) is detected optoelectronically along the capillary tube, and a flow curve can be drawn. Shear rate and shear stress are calculated from this curve by a computer program. The viscosity values are determined as a function of these parameters according to Casson's principle. For the presentation of our results, apparent whole blood viscosity values calculated at 90 1/s shear rate are given [21–24]. Fibrinogen was determined by using Clauss method. All measurements were carried out at room temperature (22 ± 1 ^{o}C) within three hours after venepuncture.

Table 1. The circulatory index (CRI)

$$\text{CRI} = \frac{\text{Cardiac index (CI) in } ml/s/m^2}{\text{Whole blood viscosity (WBV) at 90 1/s in } mPas\ (mNs/m^2)} \quad \frac{l}{Ns^2}$$

Cardiac output and cardiac index were measured by impedance cardiography (ASK, Hungary) and circulatory index was calculated according to the formula in Table 1 [7,10,22,24].

Data were evaluated by conventional (means, S.D., S.E.M., Student's tests) statistical methods on IBM computer.

3 Results

Hematocrit level of IHD patients was significantly higher than that of normal controls ($p < 0.05$-0.01). Patients with multivessel disease had significantly elevated hematocrit comparing to the other IHD groups ($p < 0.02$) (Figure 2).

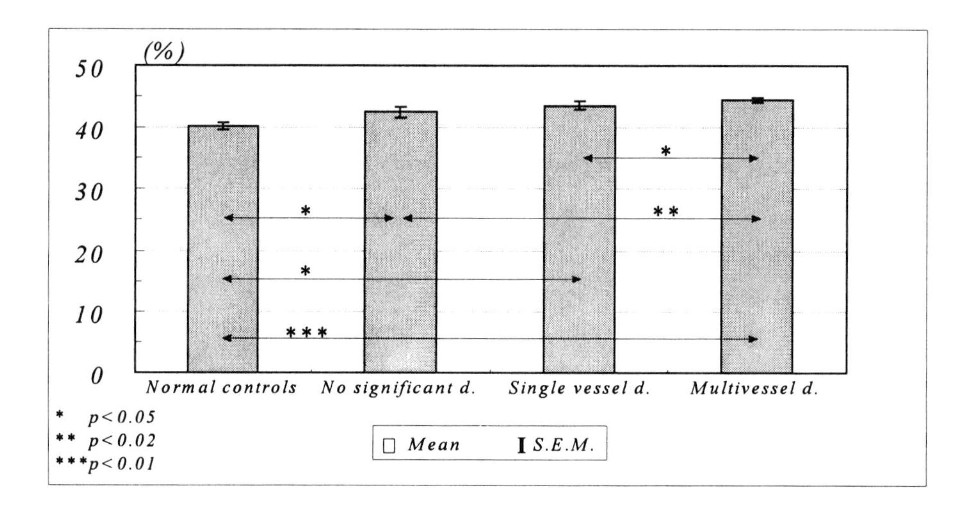

Figure 2. Hematocrit levels

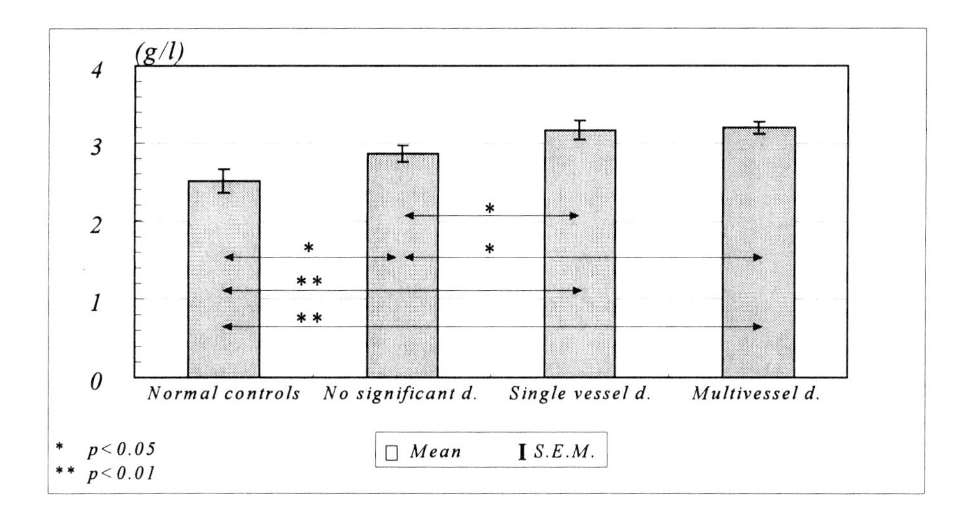

Figure 3. Plasma fibrinogen levels

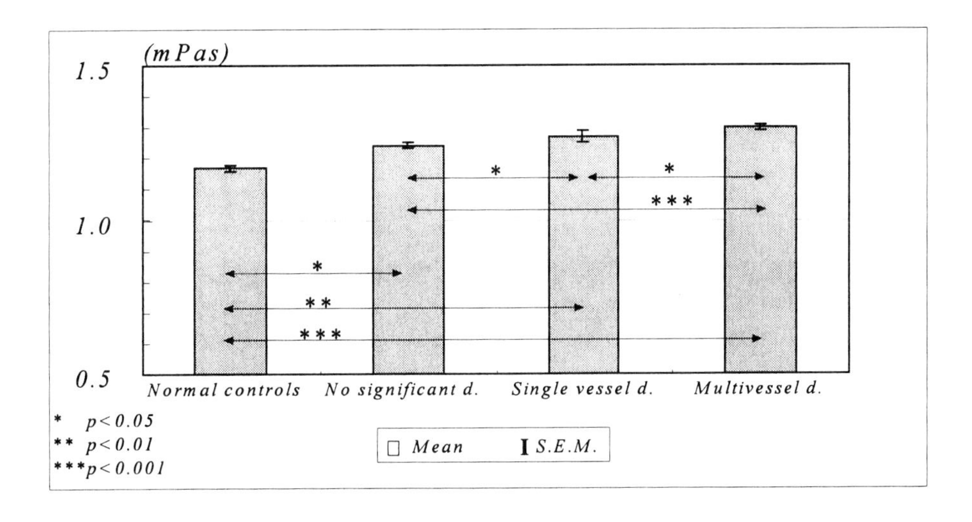

Figure 4. Plasma viscosities

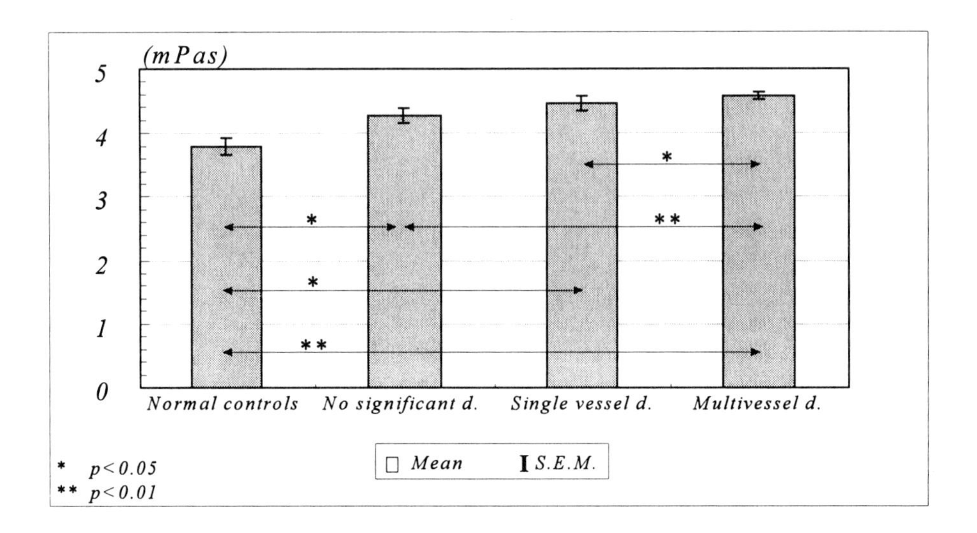

Figure 5. Whole blood viscosities

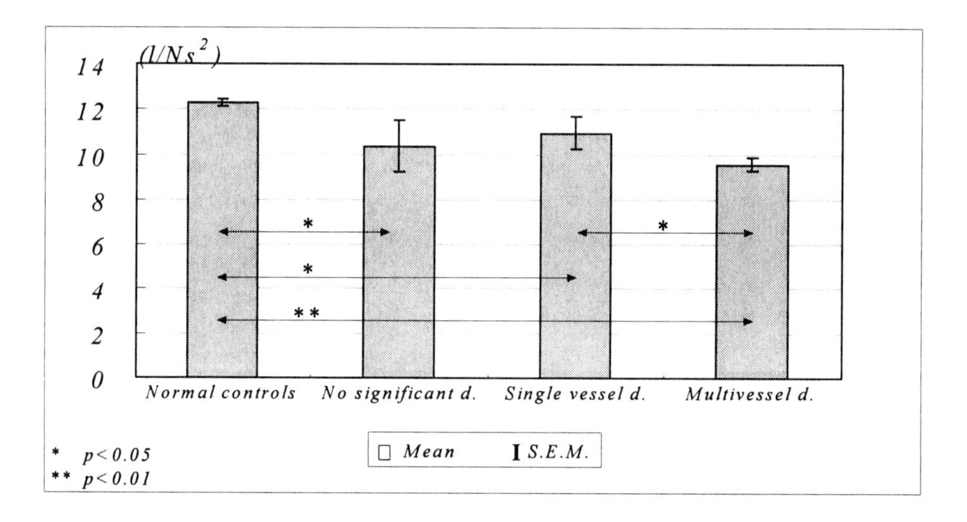

Figure 6. CRI values

Plasma fibrinogen level of normals was significantly lower than that of patients ($p < 0.05$ between G-0 and G-1, $p < 0.01$ between G-0 and G-2, and between G-0 and G-3). Plasma fibrinogen also showed a significant elevation in the single and multivessel disease groups comparing to patients without significant coronary artery disease ($p < 0.05$) (Figure 3).

Plasma viscosity was increased in IHD patients ($p < 0.05$-0.001), and in G-2 and in G-3 higher values were found comparing to G-1 ($p < 0.05$ and 0.001) (Figure 4).

Whole blood viscosity was elevated in IHD patients ($p < 0.05$-0.01). The multivessel disease group had higher levels of WBV than single vessel disease subjects and patients without significant lesions ($p < 0.01$). The two latter groups did not differ from each other in this parameter (Figure 5).

Among hemodynamic parameters, the cardiac index at rest did not differ statistically in normals and in IHD subjects, but CRI was significantly reduced in our patients ($p < 0.05$-0.01). In the multivessel disease group significantly lower values were found than in single vessel disease ($p < 0.05$) (Figure 6).

4 Discussion

Hemorheological parameters seem to be important risk factors in cardiovascular diseases. Coronary artery disease is a major, and both clinically and economically important part of these diseases, in which the role of impaired blood rheology is also emphasized [1,2,11,13,17].

In our latest study all the measured hemorheological parameters were higher and the circulatory index was lower in IHD patients than in healthy controls which is in concordance with the results of previous studies [11–26]. Moreover, fibrinogen and plasma viscosity levels were more increased in the single vessel and multivessel disease group (G-2, G-3) comparing to IHD patients without significant coronary artery lesions (G-1), which shows the more deteriorated rheological state of those patients. The increased fibrinogen and plasma viscosity levels can result in higher red blood cell aggregability and all these changes can cause further reduction in coronary blood flow which has already been affected by the stenotic lesions. Although the higher hematocrit value of multivessel disease patients may support the oxygen transport, but on the other hand it can worsen the coronary circulation by the elevation of whole blood viscosity.

In spite of the statistically not different central hemodynamic parameters, circulatory index was found to be significantly lower in patients with IHD. Furthermore, it was significantly decreased in multivessel disease, which can refer to the impaired circulation even in resting subjects. In our previous studies CRI was found to be significantly reduced in IHD both at rest and during exercise, and it was also proved to be a more sensitive parameter for the characterization of the state of circulation than whole blood viscosity or cardiac index alone [22].

There were smaller number of patients in G-1 and G-2. This lower rate of patients with less severely affected coronaries is unfortunately characteristic in our country, where most of the patients undergoing invasive procedures have poor coronary state. Therefore subjects in G-1 could not be classified into subgroups. This study should be continued in order to examine the hemorheological parameters in these patients with minor (not significant) lesions on coronary arteries, with coronary spasm or coronary X syndrome, because their coronary flow can be reduced and rheological alterations can cause even further reduction.

In summary, our findings indicate not only that hemorheological parameters are important risk factors of ischemic heart disease, but also that they can change along with the severity of the coronary artery disease and may play a pathophysiological role in the deterioration of this disease.

This work was supported by T-06 030/96 ETT and T25432 OTKA grants.

Bibliography

1. O.K. BASKURT, E. LEVI, S. CAGLAYAN, N. DIKMENOGLU, O. UCER, R. GUNER and S. YORUKAN. The role of hemorheologic factors in the coronary circulation. *Clin. Hemorheol.*, **11**, 1991, 121–127.

2. C. CARTER, D. MCGEE, D. REED, K. YANO and G. STEMMERMANN. Hematocrit and the risk of coronary heart disease: The Honolulu Heart Program. *Am. Heart J.*, **105**, 1983, 674–679.

3. S. CHIEN. Hemorheology in clinical medicine. *Clin. Hemorheol.*, **2**, 1982, 137–142.

4. L. DINTENFASS. Blood rheology in pathogenesis of the coronary heart diseases. *Am. Heart J.*, **77**, 1969, 139–147.

5. E. ERNST, W. KOENIG, A. MATRAI and U. KEIL. Plasma viscosity and hemoglobin in the presence of cardiovascular risk factors. *Clin. Hemorheol.*, **8**, 1988, 507–515.

6. E. ERNST. Plasma fibrinogen - an independent cardiovascular risk factor. *J. of Int. Med.*, **227**, 1990, 365–372.

7. I. HORVATH, B. MEZEY, I. JURICSKAY, A. SIMON and T. JAVOR. Data for the clinical evaluation of the impedance-cardiographic measurements. (Hungarian). *Card. Hung.*, **22**, 1993, 29–32.

8. W.B. KANNEL, R.B. D'AGOSTINO and A.J. BELANGER. Fibrinogen, cigarette smoking, and risk of cardiovascular disease: Insights from the Framingham Study. *Am. Heart J.*, **113**, 1987, 1006–1010.

9. W. KOENIG, E. ERNST and A. MATRAI. Blood rheology associated with cardiovascular risk factors and chronic cardiovascular diseases: Results of an epidemiologic cross-sectional study. *Angiology*, **39**, 1988, 986–995.

10. W.G. KUBICEK, J.M. KARNEGIS, R.P. PATTERSON, D.A. WITSOE and R.H. MATTSON. Development and evaluation of an impedance cardiac output system. *Aerospace Med.*, **37**, 1966, 1208–1212.

11. G.D.O. LOWE, M.M. DRUMMOND, A.R. LORIMER, I. HUTTON, C.D. FORBES, C.R.M. and J.C. BARBENEL. Relation between extent of coronary heart disease and blood viscosity. *Br. Med. J.*, **280**, 1980, 673–674.

12. G.D.O. LOWE, W.C.S. SMITH, H.D. TUNSTALL-PEDOE, I.K. CROMBIE, S.E. LENNIE, J. ANDERSON and J.C. BARBENEL. Cardiovascular risk and haemorheology - Results from the Scottish Heart Health Study and the MONICA Project, Glasgow. *Clin. Hemorheol.*, **8**, 1988, 517–524.

13. M. MARES, C. BERTOLO, V. TERRIBILE and A. GIROLAMI. Hemorheological study in patients with coronary artery disease. *Cardiology*, **78**, 1991, 11–16.

14. L. MOLLER and T.S. KRISTENSEN. Plasma fibrinogen and ischemic heart disease risk factors. *Arterioscler.-Thromb.*, **11**, 1991, 344–350.

15. F.J. NEUMANN, H. TILLMANNS, P. ROEBRUCK, R. ZIMMERMANN, H.M. MAUPT and W. KUBLER. Haemorrheological abnormalities in unstable angina pectoris: A relation independent of risk factor profile and angiographic severity. *Br. Heart J.*, **62**, 1989, 421–428.

16. F.J. NEUMANN, H.A. KATUS, E. HOBERG, P. ROEBRUCK, M. BRAUN, H.M. HAUPT, H. TILLMANNS and W. KUBLER. Increased plasma viscosity and erythrocyte aggregation - indicators of an unfavourable clinical outcome in patients with unstable angina pectoris. *Br. Heart J.*, **66**, 1991, 425–430.

17. C. RAINER, D.T. KAWANISHI, P.A.N. CHANDRARATNA, R.M. BAUERSACHS, C.L. REID, S.H. RAHIMTOOLA and H.J. MEISELMAN. Changes in blood rheology in patients with stable angina pectoris as a result of coronary artery disease. *Circulation*, **76**, 1987, pp. 15–20.

18. J.F. STOLTZ. Cardiovascular diseases, risk factors and hemorheological parameters. *Clin. Hemorheol.*, **3**, 1981, 257–267.

19. B.E. STRAUER, H. BOHME, W. SAMTLEBEN, S. SCHULIG and E. VOLGER. Clinical approach to disturbances in microcirculation of the heart. *Clin. Hemorheol.*, **4**, 1984, 45–66.

20. P.M. SWEETNAM, H.F. THOMAS, J.W.G. YARNELL, A.D. BESWICK, I.A. BAKER and P.C. ELWOOD. Fibrinogen, viscosity and the 10-year incidence of ischemic heart disease: The Caerphilly and Speedwell studies. *Eur. Heart J.*, **17**, 1996, 1814–1820.

21. K. TOTH, B. MEZEY, I. JURICSKAY and T. JAVOR. Pattern recognition in evaluation of haemorheological and haemodynamical measurements in the cardiological diagnostics. *Acta Med. Hung.*, **47**, 1990, 31–42.

22. K. TOTH, T. HABON, I. HORVATH, B. MEZEY, I. JURICSKAY and G.Y. MOZSIK. Hemorheological and hemodynamical parameters in patients with ischemic heart disease at rest and at peak exercise. *Clin. Hemorheol.*, **14**, 1994, 329–338.

23. K. TOTH, L. BOGAR, I. JURICSKAY, M. KELTAI, S. YUSUF, L.J. HAYWOOD and H.J. MEISELMAN. The effect of RheothRx Injection on the hemorheological parameters in patients with acute myocardial infarction (CORE trial substudy). *Clin. Hemorheol. and Microcirc.*, **17**, 1997, 117–125.

24. G. KESMARKY, K. TOTH, L. HABON, G. VAJDA and I. JURICSKAY. Hemorheological parameters in coronary artery disease. *Clin. Hemorheol. and Microcirc.*, **18**, 1998, 245–251.

25. E. VOLGER. Rheological aspects of coronary artery and coronary small blood vessel diseases (syndrome X). *Clin. Hemorheol.*, **4**, 1984, 209–221.

26. J.W. YARNELL, I.A. BAKER, P.M. SWEETNAM, D. BAINTON, J.R. O-BRIEN, P.J. WHITEHEAD and P.C. ELWOOD. Fibrinogen, viscosity, and white blood cell count are major risk factors for ischemic heart disease. The Caerphilly and Speedwell collaborative heart disease studies. *Circulation*, **83**, 1991, 836–844.

Hemorheological Alterations After Percutaneous Transluminal Coronary Angioplasty

G. Kesmarky*, K. Toth*, G. Vajda, L. Habon**, R. Halmosi*, T. Habon* and E. Roth†**

**Department of Medicine, Division of Cardiology, **Department of Surgery, Heart Center, †Department of Experimental Surgery, University Medical School of Pecs, Hungary*

Abstract

Percutaneous transluminal coronary angioplasty (PTCA) is a widely used method for the treatment of coronary artery disease. During this procedure several changes in blood flow occur due to coronary stenosis, endothelial injury, and ischemia-reperfusion caused by the balloon inflation/deflation. In our study 20 patients (mean age 58 ± 9 years) undergoing PTCA were examined. For the laboratory measurements several blood samples were taken from peripheral vein and the coronary sinus before and 30 minutes after PTCA, and from peripheral vein 1, 2, 5 days and 1, 6 months after PTCA. Among hemorheologic parameters hematocrit, plasma fibrinogen level, plasma and whole blood viscosities were measured and corrected blood viscosity value was calculated. Plasma fibrinogen level increased markedly during the first and second day after PTCA ($p < 0.001$), which was accompanied by the elevation of plasma viscosity ($p < 0.05$). Plasma fibrinogen returned to the baseline at the 1 month control visit, but there was a significant increase in its level again by the sixth month measurement. Apparent whole blood viscosity at 90 1/s showed gradually increasing levels up to one- and six-month control visits ($p < 0.01$), which can partially be explained by the elevation of hematocrit. Corrected blood viscosity was significantly elevated on the fifth day already ($p < 0.01$), and also one month later. Platelet aggregability increased significantly ($p < 0.05$) 30 minutes after the intervention. Our findings indicate that PTCA may cause significant changes in the hemorheologic parameters, which can affect the final result of this intervention.

1 Introduction

Percutaneous transluminal coronary angioplasty (PTCA) is a widely used method in the treatment of significant coronary stenoses. The high (30–40%) ratio of reocclusion and restenosis is the main challenge of this procedure. The mechanism of these processes involves platelet and leucocyte activation, thrombus formation, endothelial and smooth muscle cell proliferation, and local and systemic inflammation, but the pathogenesis is still only partially understood [1,2].

Hemorheological parameters were found to be risk factors of cardiovascular diseases in several studies [3–8]. The coronary circulation is determined by both hemodynamic and hemorheologic factors. In the severely stenosed coronaries, where shear stress can grow very high, the rheological properties of the non-newtonian blood can become of crucial importance [9–12]. These can also play a role in the deterioration of blood flow in the small vessels, and may contribute to the slower restoration of coronary circulation even after a successful procedure. The effects of coronary angioplasty on the parameters characterizing the blood flow properties and on the characteristics of erythrocytes have not been clarified.

The aim of this study was to investigate whether hemorheological parameters change during and after PTCA. There has been relatively lower number of studies which tried to follow up laboratory parameters several days and months after a coronary event or a coronary intervention, despite that vessel wall remodeling and restenosis can develop during this longer period. Therefore we determined the hemorheological parameters on several days after PTCA and at the first and sixth month's regular visit also.

2 Patients and methods

In the study twenty patients were involved (mean age: 58 ± 9 years). After coronary angiography, which revealed the site of the culprit stenosis, percutaneous transluminal coronary angioplasty was performed. Patients were on combined antianginal-antiischemic drug therapy (including nitrates, β-blockers, angiotensin converting enzyme inhibitors, calcium antagonists, lipid lowering drugs and conventional antiplatelet therapy: acetylsalicylic acid, ticlopidine) and heparin was administered as an anticoagulant during the procedure and for 24 hours after PTCA.

Coronary artery branch with the significant stenosis was dilated by balloon catheter which was inserted via the femoral artery. Coronary sinus was catherized from the femoral vein. Blood samples were drawn from both the coronary sinus and peripheral (femoral) vein before and thirty minutes after PTCA, and from peripheral (cubital) vein 1, 2, 5 days, 1 and 6 months after PTCA.

Hematocrit (Hct) was measured with microhematocrit centrifuge. Plasma and whole blood viscosities (PV, WBV) were measured on Hevimet 40 capillary viscosimeter (Hemorex). In this viscosimeter the flow of whole blood (or plasma) is detected optoelectronically along the capillary tube, and a flow curve can be drawn. Shear rate and shear stress are calculated from this curve by a computer program. Viscosity values are determined as a function of these parameters according to Casson's principle. For the presentation of our results, apparent whole blood viscosity values calculated at 90 1/s shear rate are given [12]. Whole blood viscosity at 90 1/s was corrected to 40% hematocrit ($WBV_{40\%}$) with a mathematical formula according to Mátrai et al. [13]:

$$(WBV_{40\%}/PV) = (WBV_{Hct}/PV)^\wedge(40\% /Hct).$$

Fibrinogen was determined by using the Clauss method. Platelet aggregability was measured in Micron M304 aggregometer. All measurements were carried out within three hours after venepuncture. Data were analyzed by conventional statistical methods (means, S.E.M., "t" test).

3 Results

There was a significant decrease in the values of hematocrit, plasma and whole blood viscosity right after PTCA, while corrected whole blood viscosity remained unchanged during this period (Table 1).

Hematocrit level returned to the baseline during the days after PTCA, and then an elevation at the end of the first and the sixth month could be observed (Figure 1). Plasma fibrinogen level elevated markedly during the days after PTCA with a peak value on the second day, then returned to the baseline after one month. Another significant increase could be seen after six months (Figure 2).

Plasma viscosity showed a similar trend to that of fibrinogen after PTCA, but seemed to remain higher than baseline during the following months (Figure 3). Whole blood viscosity - after the post-PTCA reduction - increased significantly by the end of the study period (Figure 4), while corrected blood viscosity showed a continuously increasing tendency with a significant elevation on day 5 and afterwards. (Figure 5).

Spontaneous platelet aggregation was significantly higher in the samples from the coronary sinus than in those from the peripheral vein. Spontaneous platelet aggregation in the periphery and epinephrine-induced aggregation from both sites were markedly elevated after PTCA. On the following days platelet aggregability decreased to the baseline. (Figures 6 and 7).

Table 1. Hemorheological parameters before and 30 minutes after PTCA

	Before PTCA		30 Minutes After PTCA	
	Vein	Cor. Sinus	Cor. Sinus	Vein
Hematocrit (%)	40.6 ± 0.8	40.6 ± 0.8	38.6 ± 0.9 ***	38.3 ± 1.0 ***
Plasma viscosity ($mPas$)	1.20 ± 0.02	1.20 ± 0.02	1.16 ± 0.02 **	1.18 ± 0.0[illegible] *
Whole blood viscosity at 90 1/s ($mPas$)	3.88 ± 0.13	3.98 ± 0.13	3.78 ± 0.15 **	3.68 ± 0.1[illegible] *
Corrected whole blood viscosity ($mPas$)	3.76 ± 0.08	3.86 ± 0.07	3.90 ± 0.13	3.82 ± 0.0[illegible]

* $p < 0.05$ ** $p < 0.01$ *** $p < 0.001$

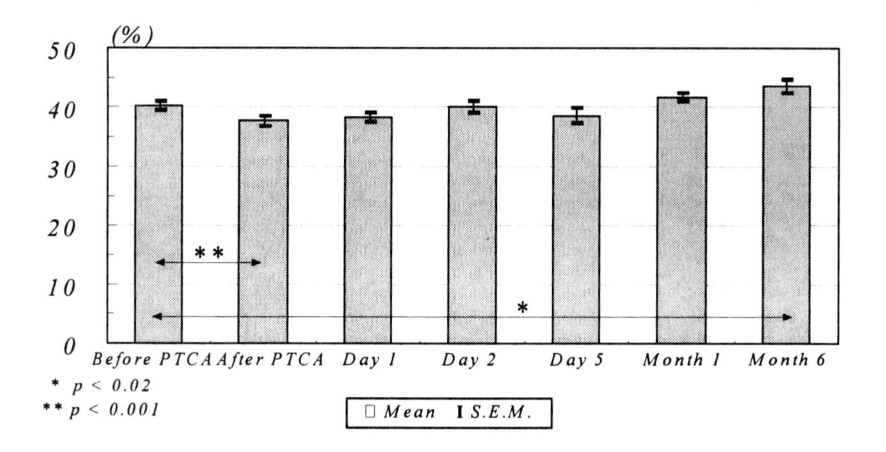

Figure 1. Hematocrit

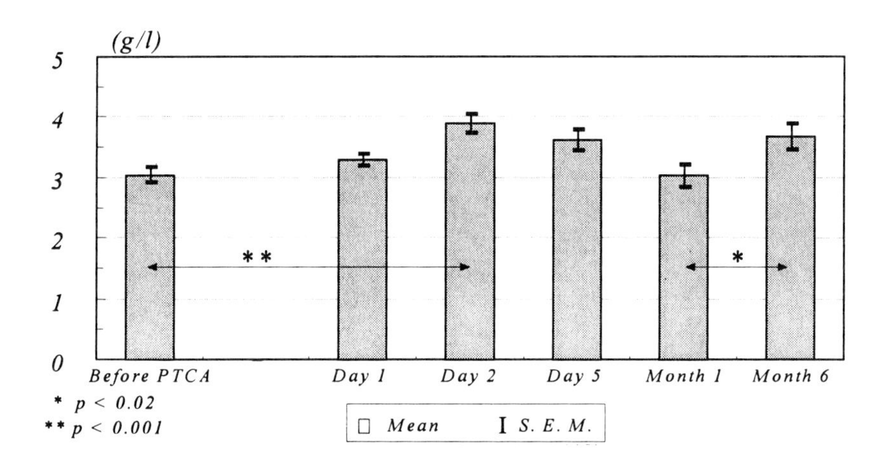

Figure 2. Plasma fibrinogen level

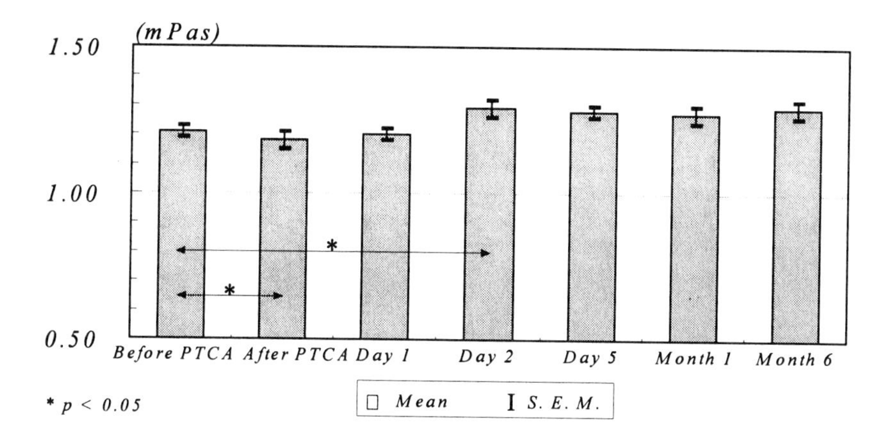

Figure 3. Plasma viscosity

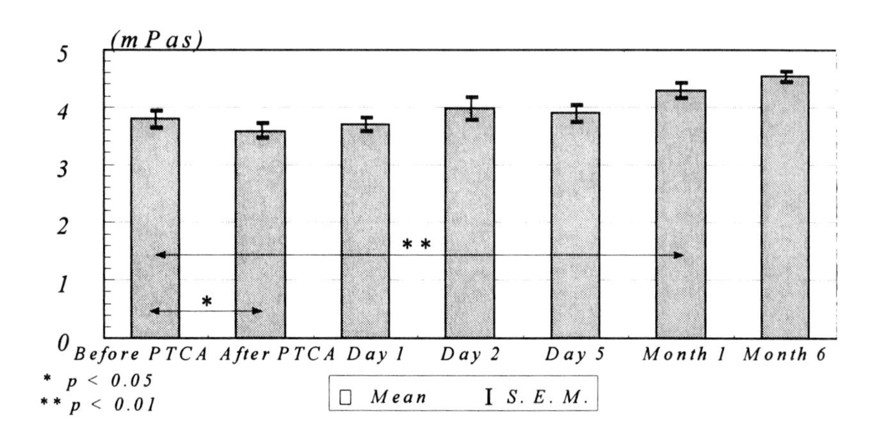

Figure 4. Apparent whole blood viscosity at 90 1/s

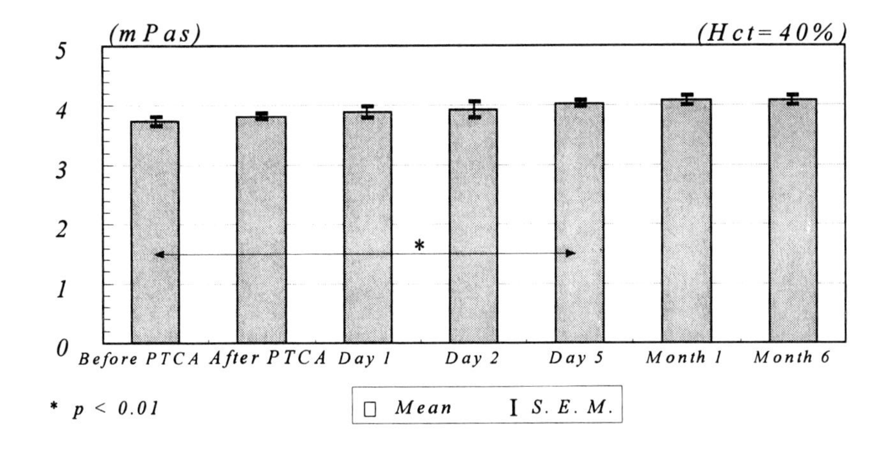

Figure 5. Corrected blood viscosity

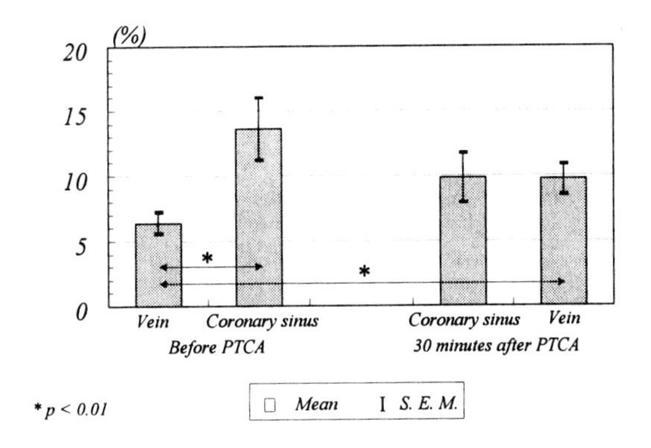

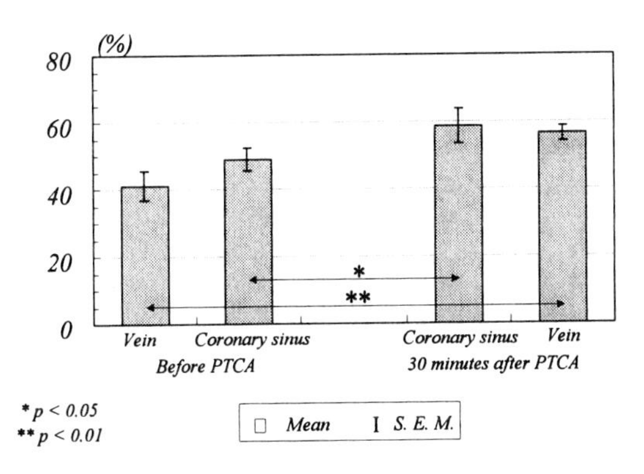

Figure 6. Spontaneous and epinephrine-induced platelet aggregation before and 30 minutes after PTCA

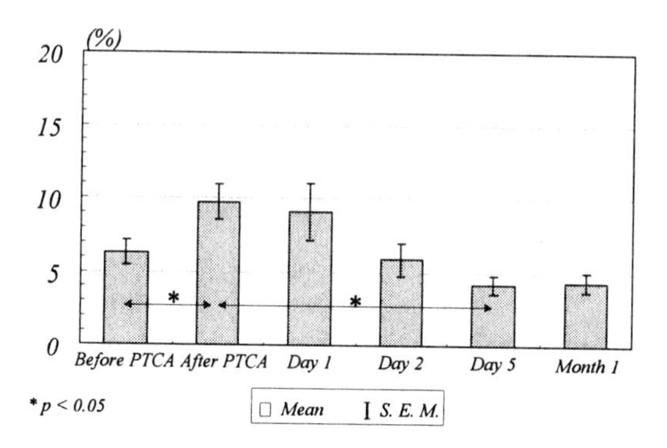

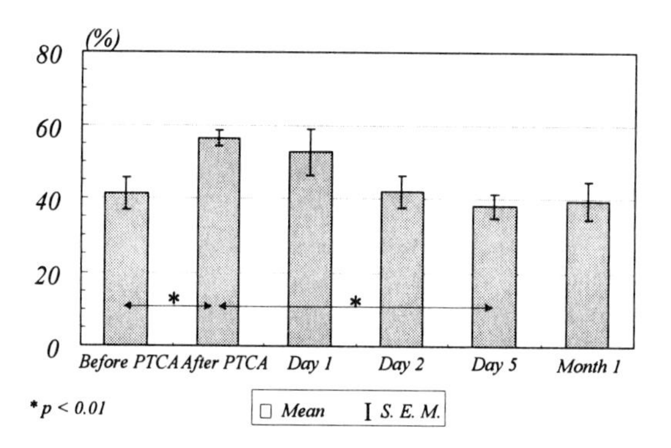

Figure 7. Spontaneous and epinephrine-induced platelet aggregation during the days and one month after PTCA

4 Discussion

Deterioration of rheological properties of blood is a known risk factor of ischemic heart disease. Elevated levels of hematocrit, fibrinogen, plasma and whole blood viscosities could be detected in patients with coronary artery disease (for example stable and unstable angina, acute myocardial infarction). Their values can correlate with the severity of coronary heart disease and the causative role of these parameters should also be considered [14–18].

In our study hematocrit, plasma and whole blood viscosity decreased significantly thirty minutes after PTCA, which could be caused by hemodilution. This could be due to the several blood samplings, some blood loss caused by the catheterization technique, the infusions and the contrast agent, respectively.

Since these changes were transient, they were not supposed to have significant advantageous effect on the coronary circulation. Corrected whole blood viscosity, a simple mathematical correction of whole blood viscosity to a standard hematocrit that allows the comparison of viscosities with different hematocrits, did not reduce after angioplasty, which supports the idea of hemodilution.

Hematocrit showed a gradual increase during the days and months following the procedure, which was significant after six months. The elevation of plasma viscosity was parallel with the changes of fibrinogen level until the second day, but then seemed to remain higher than baseline until the sixth month. Whole blood viscosity showed similar trend to that of hematocrit, although its elevation became significant by the end of the first month. The background of this deterioration in hemorheological parameters after six months is still obscure; nevertheless, these changes are similar to the results found in a previous study of our group in patients with myocardial infarction [19]. The elevation of corrected blood viscosity may refer to an impairment of the erythrocytes, which also could be affected by the procedure besides leucocytes and platelets.

Fibrinogen has been proved to be an independent risk factor of cardiovascular diseases [20]. In a recent study fibrinogen was proposed to be a risk factor of restenosis after PTCA [21]. In our study plasma fibrinogen level elevated markedly on the first two days after PTCA, then decreased to the baseline. These changes can be explained by an acute phase reaction caused by the procedure, as fibrinogen is a well-known marker of this process. Acute phase reaction occurs after organic lesion, as a part of systemic inflammatory response. In a recent study another acute phase marker, C-reactive protein was found to be increased two days after coronary angioplasty [1]. Plasma fibrinogen level is known to elevate following unstable angina and acute myocardial infarction, and higher values are associated with poorer outcome [22,23]. The elevation in fibrinogen after PTCA might increase the risk of acute ischemic coronary events, since fibrinogen plays a central role in platelet and erythrocyte aggregation and is the main determinant of plasma and whole blood viscosity [20]. More interestingly, plasma fibrinogen was elevated at the six-month sampling in correspondence with the other rheological parameters. The reason of this change is still not known (there was no signicant change in the treatment and lifestyle of the patients), so it needs further clarification whether this could elevate the risk of coronary events.

Spontaneous platelet aggregability was increased significantly in the samples from the coronary sinus before the procedure, which may refer to the activation of platelets in the diseased coronaries. Moreover, there was a further elevation in this parameter thirty minutes after PTCA, which returned to the baseline two days later. All these findings confirm that PTCA can result in platelet activation and also imply that the conventional antiplatelet combination (aspirin+ticlopidine) may be insufficient to achieve a marked reduction in platelet aggregation and thrombus formation in cases when enhanced platelet activation occurs. These findings also indicate the necessity of new antiplatelet agents (for example glycoprotein IIb/IIIa receptor blockers, etc.) in the routine clinical practice.

Our study involved a relatively low number of patients, therefore the relationship between the hemorheological parameters and restenosis could not be evaluated. Red blood cell properties were characterized indirectly by hematocrit, whole blood and corrected blood viscosity; therefore the measurement of red blood cell aggregability and deformability could reveal interesting new details.

In summary, we found significant changes in hemorheological parameters after percutaneous transluminal coronary angioplasty, which can deteriorate the coronary circulation and can affect the final outcome of this intervention. Further studies are needed to evaluate the middle and long term changes in these parameters and to clarify their role in restenosis.

This work was supported by T25432 OTKA grant.

Bibliography

1. Azar, R.R., McKay, R.G., Kiernan, F.J., Seecharran, B., Feng, Y.J., Fram, D.B., Wu, A.H.B., and Waters, D.D. (1997). Coronary angioplasty induces a systemic inflammatory response. *Am. J. Cardiol.*, **80(11)**, 1476–1478.

2. Serrano, C.V., Ramires, J.A.F., Venturinelli, M., Arie, S., D'Amico, E., Zweier, J.L., Pileggi, F., and Da Luz, P. (1997). Coronary angioplasty results in leukocyte and platelet activation with adhesion molecule expression. Evidence of inflammatory responses in coronary angioplasty. *J. Am. Coll. Cardiol.*, **29**, 1276–1283.

3. Carter, C., McGee, D., Reed, D., Yano, K., and Stemmermann, G. (1983). Hematocrit and the risk of coronary heart disease: The Honolulu Heart Program. *Am. Heart J.*, **105**, 674–679.

4. Kannel, W.B., D'Agostino, R.B., and Belanger, A.J. (1987). Fibrinogen, cigarette smoking, and risk of cardiovascular disease: Insights from the Framingham Study. *Am. Heart J.*, **113**, 1006–1010.

5. Lowe, G.D.O., Smith, W.C.S., Tunstall-Pedoe, H.D., Crombie, I.K., Lennie, S.E., Anderson, J., and Barbenel, J.C. (1988). Cardiovascular risk and haemorheology - results from the Scottish Heart Health Study and the MONICA Project, Glasgow. *Clin. Hemorheol.*, **8**, 517–524.

6. Ernst, E. (1990). Plasma fibrinogen - an independent cardiovascular risk factor. *J. Int. Med.*, **227**, 365–372.

7. Yarnell, J.W., Baker, I.A., Sweetnam, P.M., Bainton, D., O-Brien, J.R., Whitehead, P.J., and Elwood, P.C. (1991). Fibrinogen, viscosity, and white blood cell count are major risk factors for ischemic heart disease. The Caerphilly and Speedwell collaborative heart disease studies. *Circulation*, **83**, 836–844.

8. Sweetnam, P.M., Thomas, H.F., Yarnell, J.W.G., Beswick, A.D., Baker, I.A., and Elwood, P.C. (1996). Fibrinogen, viscosity and the 10-year incidence of ischemic heart disease: The Caerphilly and Speedwell Studies. *Eur. Heart J.*, **17**, 1814–1820.

9. Dintenfass, L. (1969). Blood rheology in pathogenesis of the coronary heart diseases. *Am. Heart J.*, **77**, 139–147.

10. Chien, S. (1982). Hemorheology in clinical medicine. *Clin. Hemorheol.*, **2**, 137–142.

11. Baskurt, O.K., Levi, E., Caglayan, S., Dikmenoglu, N., Ucer, O., Guner, R., and Yorukan, S. (1991). The role of hemorheologic factors in the coronary circulation. *Clin. Hemorheol.*, **11**, 121–127.

12. Toth, K., Habon, T., Horvath, I., Mezey, B., Juricskay, I., and Mozsik, G. (1994). Hemorheological and hemodynamical parameters in patients with ischemic heart disease at rest and at peak exercise. *Clin. Hemorheol.*, **14**, 329–338.

13. Matrai, A., Whittington, R.B., and Ernst, E. (1987). A simple method of estimating whole blood viscosity at standardized hematocrit. *Clin. Hemorheol.*, **7**, 261–265.

14. Lowe, G.D.O., Drummond, M.M., Lorimer, A.R., Hutton, I., Forbes, C.D., Prentice, C.R.M., and Barbenel, J.C. (1980). Relation between extent of coronary heart disease and blood viscosity. *Br. Med. J.*, **280**, 673–674.

15. Rainer, C., Kawanishi, D.T., Chandraratna, P.A.N., Bauersachs, R.M., Reid, C.L., Rahimtoola, S.H., and Meiselman, H.J. (1987). Changes in blood rheology in patients with stable angina pectoris as a result of coronary artery disease. *Circulation*, **76**, 15–20.

16. Koenig, W., Ernst, E., and Matrai, A. (1988). Blood rheology associated with cardiovascular risk factors and chronic cardiovascular diseases: Results of an epidemiologic cross-sectional study. *Angiology*, **39**, 986–995.

17. Toth, K., Mezey, B., Juricskay, I., and Javor, T. (1989). Hemorheological changes during the hospital phase of acute myocardial infarction: Sex differences? *Med. Sci. Res.*, **17**, 841–844.

18. Kesmarky, G., Toth, K., Habon, L., Vajda, G., and Juricskay, I. (1998). Hemorheological parameters in coronary artery disease. *Clin. Hemorheol. Microcirc.*, **18(4)**, 245–251.

19. Toth, K., Mezey, B., Juricskay, I., Simor, T., and Javor, T. (1990). Hemorheological changes during the first six months after acute myocardial infarction [Hungarian]. *Orv. Hetil.*, **131**, 727–730.

20. Montalescot, G., Collet, J.P., Choussat, R., and Thomas, D. (1998). Fibrinogen as a risk factor for coronary heart disease. *Eur. Heart J.*, **19**, (Supplement H), H11–H17.

21. Montalescot, G., Ankri, A., Vicaut, E., Drobinski, G., Grosgogeat, Y., and Thomas, D. (1995). Fibrinogen after coronary angioplasty as a risk factor for restenosis. *Circulation*, **92**, 31–38.

22. Neumann, F.J., Tillmanns, H., Roebruck, P., Zimmermann, R., Haupt, H.M., and Kubler, W. (1989). Haemorrheological abnormalities in unstable angina pectoris: A relation independent of risk factor profile and angiographic severity. *Br. Heart J.*, **62**, 421–428.

23. Neumann, F.J., Katus, H.A., Hoberg, E., Roebruck, P., Braun, M., Haupt, H.M., Tillmanns, H., and Kubler, W. (1991). Increased plasma viscosity and erythrocyte aggregation - indicators of an unfavourable clinical outcome in patients with unstable angina pectoris. *Br. Heart J.*, **66**, 425–430.

Adhesive Interaction Between Leukocytes and Platelets at the Vessel Wall

Gerard B. Nash, Christopher M. Kirton, Philip C.W. Stone and G. Ed Rainger

Department of Physiology, The Medical School, University of Birmingham

Abstract

Adhesive interactions between leukocytes and platelets can be demonstrated in the flowing blood or at the vessel wall, where immobilised platelets may capture leucocytes. Platelets may bind to matrix proteins exposed after damage to endothelium or to intact endothelial cells exposed to a variety of stimuli. When covering large areas, platelets support a rolling form of adhesion by all types of leukocytes, and even small numbers of platelets may briefly capture flowing cells. In the case of activated neutrophilic granulocytes, platelets can also support their immobilisation and migration. Platelets themselves may supply the neutrophil-activating signal. The molecular basis of the adhesive interactions have been well described, and from the rheological point of view, the flow rates and shear stresses at which interaction can occur have been defined in models using purified cells. However, it is not clear under what conditions such interactions can occur in whole blood, especially in the context of arterial flow where pulsatility and discontinuities in the wall may be predicted to greatly influence deposition of these cells. Animal models suggest that co-deposition can occur in arteries, and an understanding of how this comes about could significantly contribute to our understanding of atherosclerosis and of thrombotic complications, for example, associated with reconstructive surgery or angioplasty.

1 Interaction between platelets and leukocytes as a prothrombotic and proinflammatory event

Classically, leukocytes and platelets are considered to have separate functions in host defence and haemostasis respectively. However, recent studies suggest that the separation in functions may not always be clear cut, and that adhesive and biochemical interactions may promote thrombotic and inflammatory events, especially in the case of neutrophils (for example, [1] and [2] for reviews). Figure 1 gives an outline of these interactions, and makes the point that the adhesive and biochemical events are inter-related because they can promote each other. Adhesive interactions can be demonstrated in the flowing blood or at the vessel wall, where immobilised platelets may capture flowing leucocytes. The

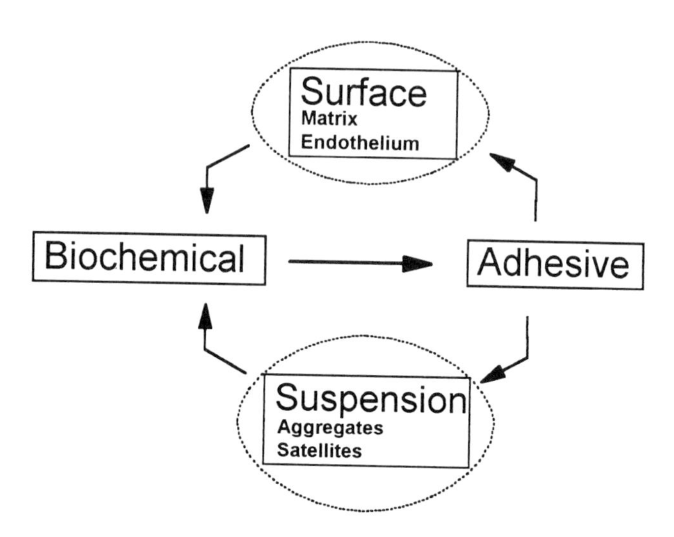

Figure 1. Summary of types of interaction between leukocytes and platelets. Adhesive interaction may occur in suspension (taking the form of aggregates of leukocytes and platelets, or of satellites formed by platelets adhered over the surface of leukocytes), or on surfaces where deposited platelets capture passing leukocytes. Adhesive interaction promotes biochemical interaction, which in turn tends to activate the cells and further enhance adhesion

ability of platelets to promote adhesion of leukocytes (particularly neutrophils) to the vessel wall is the main topic of this article. This may act to accelerate local thrombotic or inflammatory responses, both by causing accumulation of cells and by facilitating exchange of activating chemical mediators. The interaction has the potential to cause occlusion in microvessels if it occurs there, or if cell emboli break off from the walls of larger vessels. In addition, since leukocytes are not usually thought to bind to endothelium in large arteries, attachment through adherent platelets may promote deposition of leukocytes in atherosclerotic plaques.

2 Platelet deposition and subsequent adhesion of flowing neutrophils

Platelets and neutrophils utilise distinct adhesive interactions to fulfill their separate roles in haemostatic and inflammatory responses respectively. Platelets utilise glycoprotein GP1b to bind to exposed sub-endothelial matrix and then GpIIbIIIa to bind to each other in order to limit blood loss from damaged vessels [3]. Fast flowing neutrophils initially utilise specialised receptors of the selectin family which bind to fucoslylated, sialylated carbohydrate on glycoproteins to enable their capture and rolling adhesion on endothelial cells ([4] for review). On receipt of an activating signal from the surface, the neutrophils become firmly attached and migrate into tissue using their β2-integrin receptors CD11a/CD18 and CD11b/CD18. However, recent studies suggest that the platelet may form an adhesive substrate for neutrophils that fulfills many of the roles of endothelial cells (see Figure 2). Platelets deposited on denuded areas of vessel walls may allow the attachment of flowing neutrophils, while evidence is accumulating that even intact endothelium can bind platelets and then neutrophils under certain conditions.

We have previously demonstrated that a monolayer of activated platelets presents *P*-selectin that supports rolling-adhesion of flowing neutrophils in vitro [5] and also present ligand(s) that allow immobilisation of neutrophils after activation of their integrin CD11b/CD18 [6]. More recently we showed that even sparsely scattered platelets bound to collagen could capture neutrophils [7]; although the stability of initial adhesion was much reduced, neutrophils could go on to form long-lived attachments if they were activated. Activated neutrophils may also migrate over platelet monolayers [8] and through them if a chemotactic gradient is imposed [9]. Moreover, neutrophils migrating over the platelets

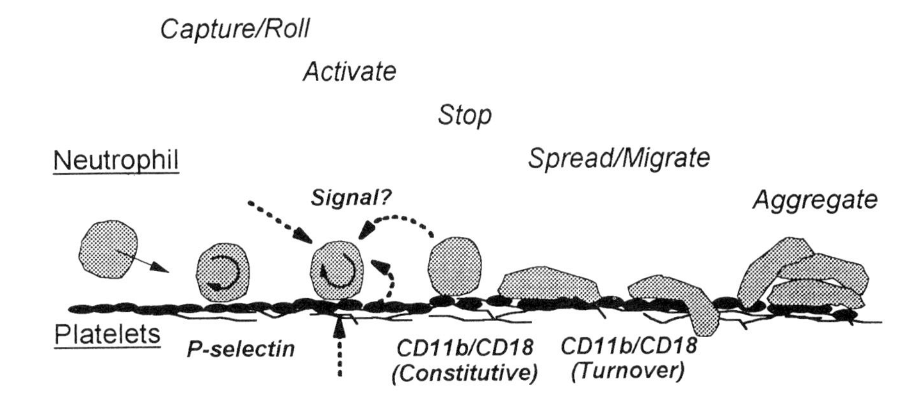

Figure 2. Adhesive interactions between immobilised activated platelets and flowing neutrophils. Initial capture and rolling adhesion is mediated by *P*-selectin on the platelets. Activating agent(s) from a number of potential sources stimulate the neutrophils so that their integrins, particularly CD11b/CD18, become activated and bind to ligands, including fibrinogen, on the platelets. The integrins "turnover" so that neutrophils can migrate while remaining attached, and may either move through the platelet monolayer or adhere to each other in aggregates

can form aggregates when they collide, leading to gradual formation of unstable groups of cells [10].

While platelets are known to attach to damaged areas of vessels, under normal circumstances they are not thought to adhere to intact endothelium. However, they may bind to endothelial cells if either type of cell is adequately stimulated ([11], for review). Platelet adhesion to endothelial monolayers has been induced *in vitro* by treatment of platelets or endothelial cells with thrombin [12–15]. Increased platelet adhesion to endothelium has also been observed *in vivo* in response to endothelial injury without denudation with a laser [16], treatment of the endothelium with IL-1 [17], IL-2 [18], calcium-ionophore A23187 [19] and agents that deplete the glycocalyx [20], and after exposure to ischaemia and reperfusion [21] or cigarette smoke [22]. Adhesion of platelets *in vivo* has been attributed to binding to endothelial P-selectin [19], or to interaction of platelet P-selectin with a receptor on peripheral lymph node endothelium [23]. *In vitro* studies with or without flow indicate that GpIIbIIIa ($\alpha II_b\beta 3$-integrin) plays an important role in attachment of activated platelets to endothelial cells [24] and [25].

Although platelets adhered to extra-cellular matrix between endothelial cells have been shown to bind flowing neutrophils [26], the ability of immobilised platelets to recruit neutrophils to an intact endothelial cell surface has not been investigated. We recently found that small numbers of platelets attached when settled onto confluent endothelial cultures, and then caused large numbers of short-lived adhesive interactions by flowing neutrophils, which could be converted to stable adhesion if the neutrophils were themselves activated [27]. In a stepwise process, P-selectin on platelets supported capture of neutrophils, which could utilise integrin receptors to become immobilised on the platelets, and move on to the endothelial cells.

Thus, if platelets adhere to the intact vessel wall *in vivo*, they may promote capture of circulating neutrophils even when the endothelial cells themselves do not present selectins. In vessels where damage occurs so that matrix is exposed, platelet deposition should also lead to capture of neutrophils. In either case, stable adhesion with onward migration of the neutrophils and formation of aggregates might ensue if the neutrophils are activated. The potential sources of such activation are worth further consideration.

3 Sources of neutrophil-activating agents which induce stable attachment

Stable attachment and migration of neutrophils requires them to receive activating signal(s). In our hands, neutrophils binding to platelets form only rolling attachments unless an external activating agent (such as bacterial peptide, activated complement or interleukin-8) is added [6–8]. In studies by others, neutrophils spontanously activate after attachment to platelets over time courses varying from seconds to tens of minutes [28–30]. The platelet-derived agent platelet-activating factor (PAF) has been implicated as the activator of neutrophils in these studies [29] and [30]. We recently noted that the ability of platelets to cause activation as well as capture of neutrophils depended, in turn, on the degree of stimulation to which the platelets had been exposed [31]. Thus after treatment with sufficient thrombin, the platelets generated an unidentified chemokine (not PAF) which caused neutrophil immobilisation. It may also be borne in mind that neutrophils themselves and endothelial cells can produce PAF and IL-8. If adherent neutrophils or adjacent endothelial cells have been sufficiently stimulated they may propogate the adhesive response by activating newly-arrived cells. In summary, fully-activated platelets may not only capture neutrophils but directly activated them. Alternatively locally or systemically released inflammatory agents may fulfill the activating requirement after platelets localise the neutrophils. There is the potential for a vicious cycle whereby cells bound to the vessel wall release agents which promote attachment of newly-arriving neutrophils.

4 Under what physical conditions can interactions occur in vivo?

The outstanding questions raised by the aforementioned studies relate to the physical conditions under which interaction between neutrophils and platelets can occur at the vessel wall *in vivo*. Thus: can neutrophils and platelets be deposited together from whole blood rather than as in all models described to date sequentially (first platelets, then neutrophils from isolated preparations); how does cell-cell interaction depend on the rheological characteristics of the blood and on the pattern of flow (for example, laminar verus turbulent and/or pulsatile flow); do conditions such as flow rate or vessel geometry in different regions of the circulation pre-dispose to adhesive interaction there? These questions require experimental studies using blood rather than purified cells, and examining a range of flow geometries and rates. As yet, no definitive answers are available. Research in this area thus offers an opportunity for those with biomengineering, modelling or cellular physiological expertise to work together on a topic relevant to cardiovascular pathology.

Bibliography

1. Bazzoni, G., Dejana, E., and Del Maschio, A. (1991). Platelet-neutrophil interactions, possible relevance in the pathogenesis of thrombosis and inflammation. *Haematolgica*, **76**, pp. 491–499.

2. Nash, G.B., Morland, C., Sheikh, S., Buttrum, S.M., and Lalor, P. (1996). Adhesion between lekocytes and platelets: Rheology, mechanisms and consequences. Progress in *Applied Microcirculation*, **22**, pp. 98–1131.

3. Williams, M.J., Du, X.P., Loftus, J.C., and Ginsberg, M.H. (1995). Platelet adhesion resceptors. *Seminars in Cell Biology*, **6**, pp. 305–314.

4. Springer, T.A. (1994). Traffic signals for lymphocyte recirculation and leukocyte emigration: The multistep paradigm. *Cell*, **76**, pp. 301–314.

5. Buttrum, S.M., Hatton, R., and Nash, G.B. (1993). Selectin mediated rolling of leukocytes on immobilized platelets. *Blood*, **82**, pp. 1165–1174.

6. Sheikh, S., and Nash, G.B. (1996). Continuous activation and deactivation of integrin CD11b/CD18 during de novo expression enables rolling neutrophils to immobilise platelets. *Blood*, **80**, pp. 1238–1246.

7. Bahra, P., and Nash, G.B. (1998). Sparsely-adherent platelets support capture and immobilisation of flowing neutrophils. *Journal of Laboratory and Clinical Medicine*, **132**, pp. 223–228.

8. Rainger, G.E., Buckey, C., Simmons, D.L., and Nash G.B. (1997). Coss-talk between cell adhesion molecules regulates the migration velocity of neutrophils. *Current Biology*, **7**, pp. 316–325.

9. Diacovo, T.G., Roth, S.J., Buccola, J.M., Bainton, D.F., and Springer, T.A. (1996). Neutrophil rolling, arrest and transmigaration across activated surface adherent platelets via sequential action of *P*-selectin and the beta-2 integrin CD11b/CD18. *Blood*, **88**, pp. 146–157.

10. Rainger, G.E., Buckley, C., Simmons, D.L., and Nash, G.B. (1998). Neutrophils rolling on immobilised platelets migrate into homotypic aggregates after activation. *Thrombosis and Haemostasis*, **79**, pp. 1177–1183.

11. Rosenblum, W.I. (1997). Platelet adhesion and aggregation without endothelial denudation or exposure of basal lamina and/or collagen. *Journal of Vascular Research*, **34**, pp. 409–417.

12. Czervionke, R.L., Hoak, J.C., and Fry, G.L. (1978). Effect of aspirin on thrombin induced adherence of platelets to cultured endothelial cells from the vessel wall. *Journal of Clinical Investigation*, **62**, pp. 841–856.

13. Chen, S., Barmatoski, S., and Barnhart, M.I. (1979). Effect of Thrombin on platelet vessel wall interactions. *Scanning Electron Microscopy*, **3**, pp. 783–789.

14. Kaplan, J.E., Moon, D.G., Weston, L.K., Minnear, F., Vecchio, P., Shepard, J., and Fenton, J. (1989). Platelets adhere to thrombin treated endothelial cells in vitro. *American Journal of Physiology*, **257**, pp. H423–433.

15. Venturini, C.M., Weston, L.K., and Kaplan, J.E. (1992). Platelet cGMP but not cAMP inhibits thrombin induced platelet adhesion to pulmonary endothelium. *American Journal of Physiology*, **263**, pp. H606–H612.

16. Povlishock, J.T., and Rosenblum, W.I. (1987). Injury of brain microvessels with a helium-neon laser and Evans blue can elicit local platelet aggregation without endothelial denudation. *Archives of Patholology and Laboratory Medicine 111*, pp. 415–421.

17. Narworth, P.P., Handley, D.A., Esmon, C.T., and Stern, D.M. (1986). Interleukin 1 induces endothelial cell procoagulant while suppressing cell-surface activity. *Proceedings of the National Acadademy of Sciences (USA)*, **83**, pp. 3460–3464.

18. Edwards, M.J., Miller, F.N., Sims, D.E., Abney, D.L., Schuschke, D.K., and Coro, T.S. (1992). Interleukin 2 acutely induces platelet and neutrophil-endothelial adherence and macromolecular leakage. *Cancer Research*, **52**, pp. 3425–3231.

19. Frenette, P.S., Johnson, R.C., Hynes, R.O., and Wagner, D.D. (1992). Platelets roll on stimulated endothelium in vivo: An interaction mediated by endothelial *P*-selectin. *Proceedings of the National Acadademy of Sciences (USA)*, **92**, pp. 7450–7454.

20. Shanberge, J.N., Kajiwara, Quattrociocchi-Longe, T. (1994). Effect of aspirin and iloprost on adhesion of platelets to intact endothelium in vivo. *Journal of Laboratory and Clinical Medicine*, **125**, pp. 96–101.

21. Massberg, S., Enders, G., Leiderer, R., Eisenmenger, S., Vestweber, D., Krombach, F., and Messmer, K. (1998). Platelet-endothelial cell interactions during ischemia/reperfusion: The role of *P*-selectin. **Blood**, **92**, pp. 507–515.

22. Lehr, H.A., Frei, B., and Arfors, K.E. (1994). Vitamin-C prevents cigarette smoke-induced leukocyte aggregation and adhesion to endothelium in vivo. *Proceedings of the National Acadademy of Sciences (USA)*, **91**, pp. 7688–7693.

23. Diacovo, T.G., Puri, K.P., Warnock, R.A., and Springer, T.A., and von Adrian, U.H. (1996). Platelet mediated lymphocyte delivery to high endothelial venules. *Science*, **273**, pp. 252–255.

24. Bombeli, T., Schwartz, B.R., and Harlan, J.M. (1998). Adhesion of activated platelets to endothelial cells: Evidence for a GPIIb-IIIa dependent bridging mechanism and roles for endothelial intercellular adhesion molecule 1 (ICAM 1), $\alpha v\beta 3$ intergrin, and GP1ba. *Journal of Experimental Medicine*, **187**, pp. 329–339.

25. Reininger, A.J., Korndorfer, M.A., and Wurzinger, L.J. (1998). Adhesion of ADP-activated platelets to intact endothelium under stagnation point flow in vitro is mediated by the integrin αIIbβ3. *Thrombosis and Haemostasis*, **79**, pp. 998–1003.

26. Kuijper, P.H., Gallardo Torres, H.F., van der Linden, J.A., Lammers, J.W., Sixma, J.J., Koenderman, L., and Zwaginga, J.J. (1996). Platelet dependent primary haemostasis promotes selectin and integrin mediated neutrophil adhesion to damaged endothelium under flow conditions. *Blood*, **87**, pp. 3271–3281.

27. Kirton, C.M., and Nash, G.B. (1997). Platelets can promote the capture and immobilisation of flowing neutrophils on a confluent endothelial cell surface. *Thrombosis and Haemostasis Supplement*, p. 600, (abstract).

28. Yeo, E.L., Sheppard, J.A.I., and Feuerstein, I.A. (1994). Role of *P*-selectin and leukocyte activation in polymorphonuclear cell adhesion to surface adherent activated platelets under physiologic shear conditions (an injury vessel wall model). *Blood*, **83**, pp. 2498–2507.

29. Weber, C, and Springer, T.A. (1997). Neutrophil accumulation on activated, surface-adherent platelets in flow is mediated by interaction of Mac-1 with fibrinogen bound to αIIbβ3 and stimulated by platelet-activating factor. *Journal of Clinical Investigation*, **100**, pp. 2085–2093.

30. Ostrovsky, L., King, A.J., Bond, S., Mitchell, D., Lorant, D.E., Zimmerman, G.A., Larsen, R., Niu, X.F., and Kubes, P. (1998). A juxtacrine mechanism for neutrophil adhesion on platelets involves platelet-activating factor and a selectin-dependent activation process. *Blood*, **91**, pp. 3028–3036.

31. Stone, P.C.W., and Nash, G.B. (1999). Conditions under which immobilised platelets activate as well as capture flowing neutrophils. *British Journal of Haematology*. In press.

Rouleaux Formation; its Causes and Consequences

M.W. Rampling

Imperial College School of Medicine, London

Abstract

Human erythrocytes in whole blood have a natural tendency to adhere together to form aggregates with characteristic morphology known as rouleaux. The ultimate cause of this phenomenon is the presence in the plasma of large plasma proteins, especially fibrinogen. Because the concentrations of these proteins increase in association with many clinical conditions, the degree of aggregation also increases in these conditions, examples of which are diabetes, hypertension, post-surgery and pregnancy. From a rheological point of view the interest in rouleaux formation is largely due to its being the major factor responsible for the remarkable shear dependence of blood viscosity. While much is already known about rouleaux formation, there are two areas that are still little understood.

The first of these arises from the relatively recent realisation that it is not just extra-cellular factors (for example plasma proteins) that cause differences in the degree of rouleaux formation between subjects, but also factors intrinsic to the cells themselves. Thus, when suspended in identical aggregant solutions, cells from different human subjects aggregate to different degrees and, even within an individual, cells show remarkable variation in their inherent potential for aggregation. The reasons are still unclear.

The second is the haemodynamic influence of rouleaux formation. There is a common assumption that because enhanced rouleaux formation is a characteristic of many diseased conditions and leads to higher viscosities (when measured in couette viscometers) that it is haemodynamically disadvantageous. Hard data in the area is, however, hard to come by and frequently ambiguous.

1 Introduction

Rouleaux formation is the loose aggregation of red cells that occurs in most human blood when left for a time in stasis. The aggregates have a characteristic morphology with the cells adherent face-to-face to produce columns of cells similar to piles of coins. The forces leading to aggregation are weak, so if a sample of normal blood is put into a couette viscometer and subjected to increasing shear rate the aggregates progressively break up until at excess of about 10 s^{-1} the aggregates are generally monodispersed.

A variety of optical techniques have shown that the aggregates take a considerable time to form. Typically Chen et al. [1] observed the rate of aggregation of samples of cells in stasis immediately after they had been disrupted by high shear. It was frequently found that tens of seconds were needed before equilibrium levels of aggregation were achieved.

These sort of data are all produced in vitro, and would be of little interest unless rouleaux were present in the living circulation, and there is compelling evidence that they do exist in vivo. Red cell aggregates can be easily seen flowing in the microcirculation of the human nail bed and in the retina. In animal studies they have been seen in less accessible areas of the microcirculation, for example in the mesentery of the rat [2]. It is more difficult to see rouleaux formation in large vessels but ultrasonic backscattering techniques developed by the groups of Boynard [3], Cloutier [4] and others have shown it to be present in blood flowing in vivo in both the venous and arterial sides of the circulation.

The major haemorheological interest in rouleaux formation lies in the fact that it gives blood complex rheological characteristics, at least when investigated in vitro [5]. Thus it leads to viscoelasticity and thixotropy and gives blood viscosity time-dependency, memory and hysteresis effects. It also endows blood with substantial shear-thinning characteristics as illustrated in Figure 1, which shows the dramatic fall in viscosity as the shear rate increases, even in a normal sample of human blood. The most rapid, early fall at shear rates up to about

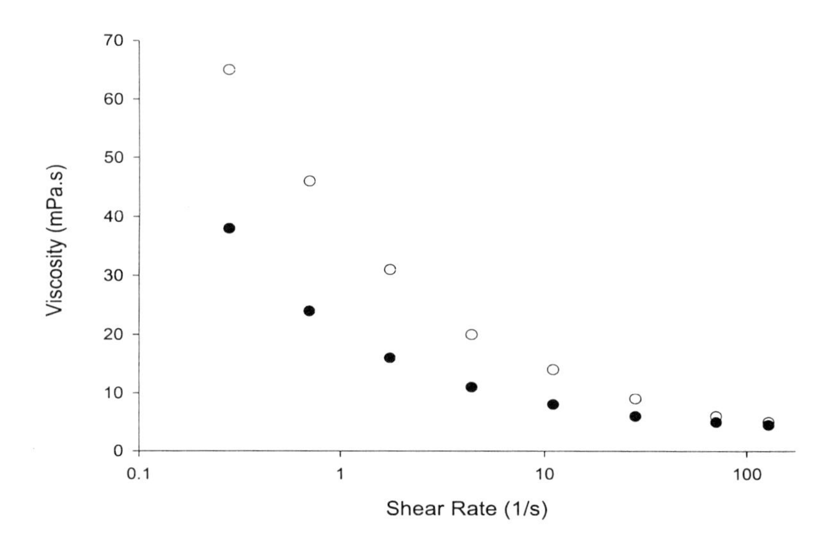

Figure 1. The viscosity:shear rate dependence of a sample of normal blood and one from a diabetic subject (closed and open symbols respectively

10 s^{-1} is usually interpreted as due to the progressive break-up of the cellular aggregates by the rising shear forces, and the later much slower rate of fall in viscosity as due to the later deformation of the cells by the increasing shear forces.

However what really matters is not just that the rheological properties of blood are complex (that would be of little more than academic interest), but that they become disturbed in association with many clinical conditions. For example, Figure 1 illustrates the fact that blood from diabetics frequently exhibits enhanced shear thinning. However, rheological disturbances are found in many clinical states, for example during pregnancy, hypertension, post surgery etc. [5]. This is of considerable clinical significance because if the rheological changes that are seen in vitro also exist in vivo, then it can be expected that they will have haemodynamic effects in vivo and that they could influence the course of the relevant clinical conditions.

2 Causes of rouleaux formation

Figure 2 shows a viscosity:shear rate plot for suspensions of cells from the same blood sample at the same haematocrit but either in autologous plasma or serum

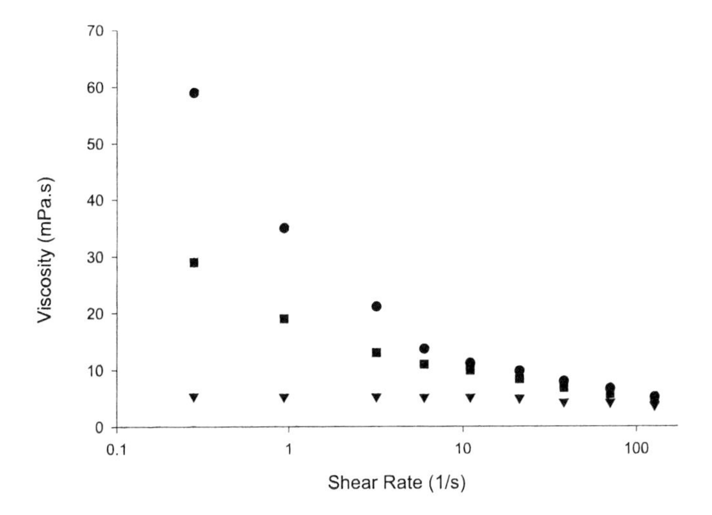

Figure 2. The viscosity:shear rate dependence of suspensions of red cells at a common haematocrit in autologous plasma, serum or in physiological protein-free buffer (circles, squares and triangles respectively)

or in protein-free buffer. It shows that removing fibrinogen, which is the only significant change in going from plasma to serum, has a very large effect on shear thinning in spite of the fact that fibrinogen makes up only about 5% of the total plasma protein concentration of a normal blood sample. All the other plasma proteins taken together are only about as effective as fibrinogen alone. The protein-free system clearly shows no shear thinning at all at low shear rate. Microscopy confirms what would be the obvious interpretation of these viscometric data, i.e. that aggregation levels in the suspensions fall progressively in moving from plasma to serum to buffer (the last of which shows no aggregation at all). This illustrates the well established fact that in normal blood, red cell aggregation is due mainly to fibrinogen but that other plasma proteins are involved as well, mainly IgG, IgM and α_2-macroglobulin [6].

While these plasma proteins are the natural causes of rouleaux formation, there are a variety of other materials that produce a similar form of aggregation when added to red cell suspensions; these include heparin, PVP and high molecular weight dextran [7]. These considerations raise three important points:-

2.1

A common feature of the aggregant molecules is that they are all large or very elongated or both, for example fibrinogen has a molecular weight of some 360,000 and an axial ratio of some 10:1 [8]. Furthermore, some are available in a variety of molecular weights, for example dextran, and the bigger they are the more effective they are as aggregating agents [9].

2.2

A very disparate feature, however, is the nature of the various molecules involved: the proteins are complex chains of different polypeptides with various charges scattered about them, while the other materials have relatively simple repetitive structures, some being negatively charged (heparin) and others neutral molecules (dextran and PVP).

2.3

Fibrinogen and the other plasma proteins that stimulate rouleaux formation are all increased dramatically in a range of clinical conditions, for example diabetes, hypertension, infection, tissue trauma and pregnancy, and this is the main reason that these conditions are associated with increases in the intensity of rouleaux formation.

3 Concentration dependence

The way in which aggregation of red cells varies as a function of the concentration of dextran is fairly well established [9] and is indicated in Figure 3. There may

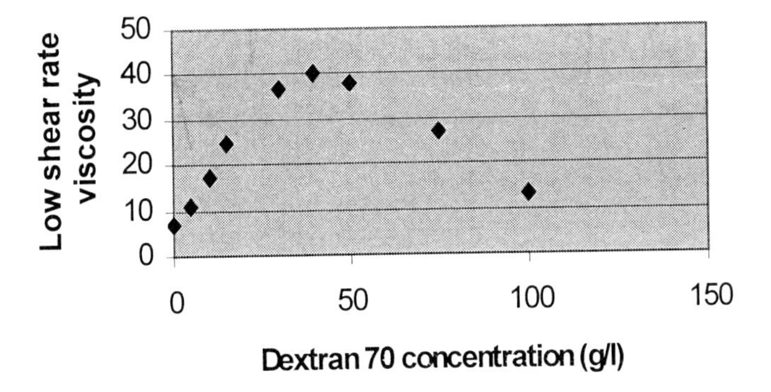

Figure 3. The variation of red cell aggregation (low shear rate viscosity as the index) with concentration of dextran 70 as aggregant

be a small threshold below which no aggregation occurs, but above the threshold aggregation increases to a peak and then it falls. Whittingstall et al. [7] have shown similar concentration dependence for heparin and PVP. The proteins also cause similar effects, but it is difficult to get enough into solution to pass the peak and get into the falling phase.

What has been discussed so far is reasonably well established but there are two areas that are far less clear; those of the essential mechanism causing aggregation and the cellular factors that are relevant.

4 Essential mechanism

The basic mechanism leading to rouleaux formation is still debated, but the most commonly favoured hypothesis is that based on the concept of molecular cross-bridging by the aggregant molecules [9]. It is assumed that the long molecules adhere to the membranes of adjacent cells over a sufficient distance that the natural, negative charge-based repulsion of the membranes is negligible. This is compatible with the fact that all the known rouleaux forming molecules are large. The problem with this hypothesis, however, is that it is difficult to understand how such varied molecules, as those known to cause rouleaux formation, all adhere with similar force. There is a natural tendency to expect adhesive phenomena to be more specific. Thus another hypothesis has arisen based on the principle of a depletion layer [10]. In this hypothesis it is assumed that the exclusion by steric hindrance of the large aggregant molecules from the glycocalyx regions on the surfaces of the cells gives rise to a depletion layer and to osmotic forces pushing the cells together. This has the distinct advantage, over the cross-bridging hypothesis, that it requires a much less specific phenomenon to be causative and is more compatible with the very different natures of the aggregant molecules.

Differentiating between these hypotheses has proven difficult and it is quite possible that both mechanisms are involved to differing degrees depending on the aggregants involved.

5 Cellular factors

Another area where much still remains to be understood is that of the cellular factors that influence aggregation potential. Meiselman was probably the first person seriously to raise this concept. He showed that cells from different individuals suspended under identical conditions in dextran solutions would aggregate to substantially different degree [11]. This implied that the cells from these different people were inherently different in their rouleaugenic potential. He was also able to show that the cells from an individual, if separated as a function of age, would aggregate to different degrees. Thus the old cells aggregated the most and the youngest the least, under otherwise identical conditions of dextran-induced aggregation [11]. These results show that not only does the average cell population from different people differ, but even within the individual cells differ in their aggregating potential. Rampling et al. [12] have shown similar results for fibrinogen-induced aggregation.

While these data are very interesting, they are proving extremely difficult to explain. The effect of age on the cell's ability to aggregate does not seem to be due to age-dependent differences in surface charge, deformabilty or size between the cells [11]. It may be due to differences in their ability to adsorb the aggregants, perhaps brought on by some form of conditioning during the cell's life in the circulation. It is more difficult to see how it can be due to a more pronounced depletion layer in the older cells as, if anything, the glycocalyx is likely to be reduced with age of the cell as a result of non-specific digestion of the coat. As to the cause of the differences between the cell populations from different people, there is almost no information on this.

To try to shed more light on these phenomena we and others have looked at mammals other than the human, and have recently shown that horse red cells aggregate much more, and cattle much less, than those from the human whether in fibrinogen, dextran, heparin or PVP suspensions. Again there does not seem to be any relation between rouleaugenicity and red cell size, surface charge or deformability differences between the mammals.

Another avenue of research that we, and others, have tried is to digest the red cell membrane surface with enzymes and to look at the effects on aggregation potential. An obvious enzyme to use is neuraminidase, which simply removes sialic acid and, thus, most of the negative charge from the red cell. It can be seen from Figure 4 that both fibrinogen- and dextran-induced aggregation are enhanced by this treatment, as might be expected because removal of the sialic

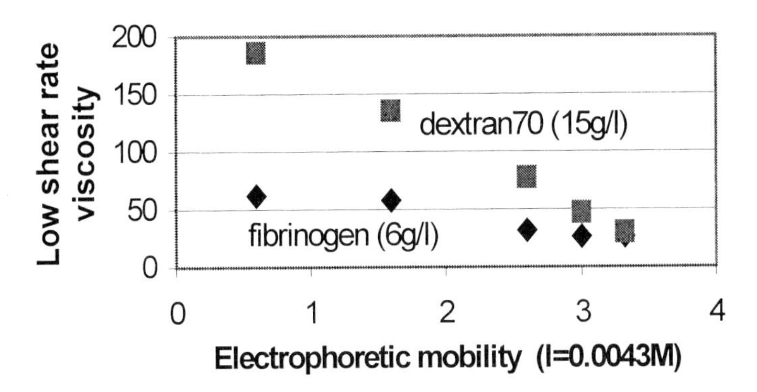

Figure 4. Low shear rate viscosity (as an index of red cell aggregation) plotted against electrophoretic mobility. The cells were suspended in physiological saline containing either dextran 70 or fibrinogen. The points (going right to left) correspond to cells which have been washed or digested to equilibrium with chymotrypsin, trypsin , bromelain or neuraminidase respectively [13]

acid dramatically reduces surface charge and, therefore, membrane-to-membrane repulsion. What was less predictable was the effect of digesting the cells with a variety of proteolytic enzymes that remove different parts of the proteinatious glycocalyx coat surrounding the membrane. It was found that any proteolytic digestion attempted led to increases in dextran-induced aggregation and with the effectiveness increasing in going from chymotrypsin to trypsin to bromelain, but with neuraminidase digestion being most effective of all. Fibrinogen-induced aggregation showed a similar pattern but with some significant differences. First, the changes were much less pronounced. Second, chymotrysin has no effect on fibrinogen-induced aggregation at all and there was no difference between the effects of bromelain and neuraminidase. An important point is that all of the enzymes led to loss of surface charge density as judged by changes in electophoretic cellular mobility, with the effectiveness of the enzymes following the same pattern as above, i.e. chymotrypsin to trypsin to bromelain to neuraminidase [13]. What this means with regard to mechanisms is not certain, but it strongly suggests that the two molecules, dextran and fibrinogen, have differing contributions from cross-bridging and depletion-layer activity bonding the cells together; in particular, that for dextran the loss of charge is predominant, while for fibrinogen loss of charge is partly counteracted by the reduction in the influence of the depletion-layer and/or the cross-bridging.

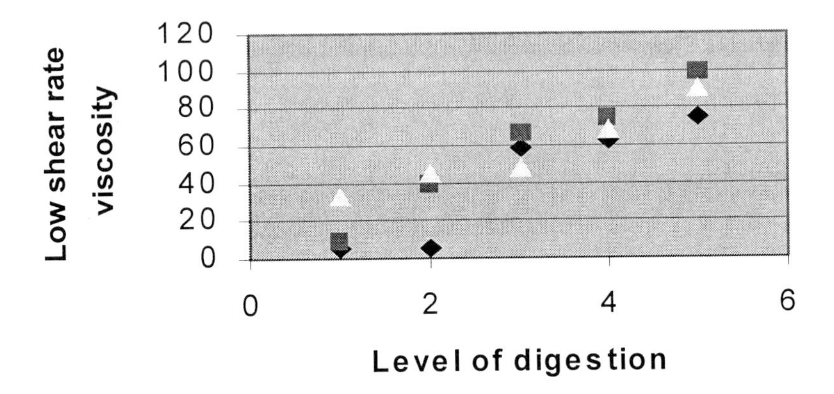

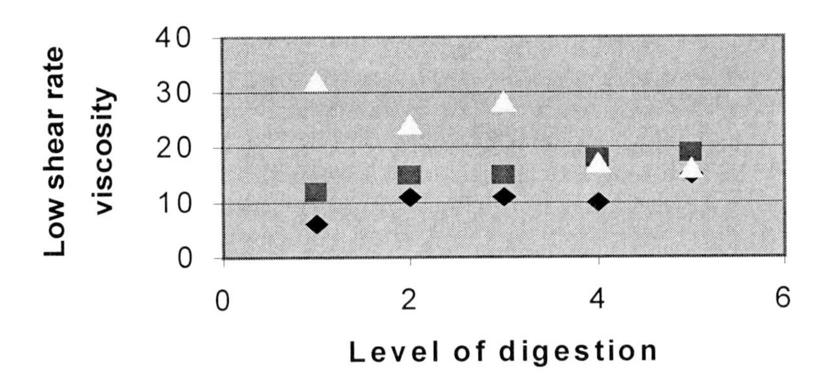

Figure 5. The effect of progressive digestion with bromelain on red cell aggregation (low shear rate viscosity as the index). The cells were suspended in $15g/l$ dextran 70 in the upper panel and in $6g/l$ fibrinogen in the lower. The data are for horse, human and cattle red cells (triangle, square and diamond respectively)

In other studies, we have tried the effects of enzyme digestion on red blood cells from other mammals, in particular horse and cattle which respectively rouleaux much more and much less than the human. What was done was to compare the effects of progressive bromelain digestion between the mammals (Figure 5). The effect on dextran-induced aggregation was similar across the three mammals, with progressive increase in aggregation with increasing digestion. However, the fibrinogen case was different. The effects on human and cow were similar with aggregation increasing with digestion, although much less than for the dextran, but the main difference was that the horse cells reduced rather than increased their aggregation as digestion progressed. Once again it is difficult to interpret these data, but it does again suggest removal of binding sites for fibrinogen being counteracted by the effect of cellular charge reduction, while for dextran only charge reduction is relevant.

Yet another avenue of study is the comparison of the membrane compositions of mammals that show very different levels of rouleaux formation. Thus Baskurt et al. [14] have shown substantial protein and carbohydrate differences between the membranes of horse and human red cell.

6 Significance of rouleaux formation

There has been a general view in the past that excessive rouleaux formation is haemodynamically disadvantageous. The reasons put forward for this have mainly been dependent on the fact that rouleaux formation leads to increased blood viscosity and to a yield stress and the consequent assumption that aggregated red cells would plug small blood vessels [6]. Furthermore, there is some in vivo data which would seem to confirm these ideas [15]. On the other hand, there are data suggesting that the enhanced axial accumulation, under the influence of rouleaux formation, of flowing cells can be haemodynamically useful and also increases leukocyte and platelet interaction with the vessel endothelium, i.e. rouleaux formation has biological advantages [16]. Thus the area is confused. However the confusion is all the greater when it is realised that there are enormous differences in the level of rouleaux formation across the mammalian kingdom [17], and it is not clear what relevance this has to the animals involved.

7 Future needs

What has been given so far is a brief summary of our understanding of rouleaux formation. However, there are important gaps that need to be plugged:-

7.1

Much more needs to be discovered about the cellular factors that lead to variation in rouleaugenicity, and whether these factors vary with clinical conditions.

7.2

More information is needed on the level of rouleaux formation that actually exists in vivo in order to judge the degree to which rouleaux-dependent haemorheological phenomena, such as shear thinning, thixotropy, axial streaming, yield stress and viscoelasticity, occur in vivo.

7.3

More studies are needed on how rouleaux-dependent haemorheological phenomena affect vascular performance and haemodynamics in vivo. For example, data does exist suggesting that significant aggregation occurs in vivo so, does the shear dependent nature of blood viscosity lead to significant blunting of flow profiles, and is this further disturbed in disease where red cell aggregation is high? Then again in the arteries where complex, pulsatile flow obtains, do thixotropy and viscoelasticity significantly come into play?

7.4

Many of the above considerations apply to large vessels, but there is also much work needed on micro-vascular flow to answer a variety of outstanding questions. For example, does red cell aggregation lead to plugging and deleterious changes, or is it advantageous in that it leads to axial flow, and allows the leukocytes and platelets to interact with the vascular surface more easily, especially in disease where aggregation is high? And again, what effect does rouleaux formation have on wall shear stress? This is important with the current interest in vascular stress detectors.

Bibliography

1. Chen, S., Barshtein, G., Gavish, B., Mahler, Y., and Yedgar, S. (1994). Monitoring of red blood cell aggregability in a flow-chamber by computerized image analysis. *Clin. Hemorheol.*, **14**, pp. 497–508.

2. Lipowski, H.H. (1998). The mechanics of leukocyte-endothelium interactions. *20th European Conference on Microcirculation*, **PO28**.

3. Lardoux, H., Boynard, M., Isnard, R., Guillet, R., Taleb, L., and Thomas, D. (1995). Left atrial spontaneous echo contrast and red cell aggregation. *Biorheology*, **32**, pp. 123–124.

4. Cloutier, G., Weng, X., Roederer, G.O., Allard, L., Tardif, F., and Raymond, B. (1997). Differences in the erythrocyte aggregation level between veins and arteries of normolipemic and hyperlipidemic individuals. *Ultrasound in Med. and Biol.*, **23**, pp. 1383–1393.

5. Rampling, M.W. (1988). Red cell aggregation and yield stress. *Clinical Blood Rheology*, **I**, Editor: G.D.O. Lowe, CRC Press, Boca Raton, pp. 45–68.

6. Lowe, G.D.O. (1988). *Clinical Blood Rheology*, **II**, CRC Press, Boca Raton.

7. Whittingstall, P., Toth, K., Wenby, R.B., and Meiselman, H.J. (1994). Cellular factors in RBC aggregation:effects of autologous plasma and various polymers. *Hemorheologie et Aggregation Erythrocytaire*, Editor: J.F. Stoltz, Editions Medicales Internationales, pp. 21–30.

8. Doolittle, R.F. (1983). The structure and evolution of vertebrate fibrinogen. *Ann. N.Y. Acad. Sci.*, **408**, pp. 13–26.

9. Chien, S., and Jan, K-M. (1973). Ultrastructural basis of the mechanism of rouleaux formation. *Microvasc. Res.*, **5**, pp. 155–166.

10. Evan, E., and Needham, D. (1988). Intrinsic colloidal attraction/repulsion between lipid bilayers and strong attraction induced by non-absorbing polymers. *Molecular mechanisms of membrane fusion*, Editors: S. Oki, D. Doyle, T.D. Flanagan, S.W. Hui and E. Mayhew, Plenum Press, New York, pp. 83–99.

11. Meiselman, H.J. (1993). Red blood cell role in RBC aggregation: 1963-1993 and beyond. *Clin. Hemorheol.*, **13**, pp. 575–592.

12. Rampling, M.W., and Whittingstall, P. (1986). The effect of cell age on the aggregability of erythrocytes. *Sixth International Congress on Biorheology*, **P29**.

13. Pearson, M. (1996). An investigation into the mechanisms of rouleaux formation and the development of improved techniques for its quantitation. *Ph.D. Thesis*, University of London.

14. Baskurt, O.K., Farley, R.A., and Meiselman, H.J (1997). Erythrocyte aggregation tendency and cellular properties in horse, human, and rat: A comparative study. *Am. J. Physiol.*, **273**, pp. H2604–H2612.

15. Cabel, M., Meiselman, H.J., Popel, A.S., and Johnson, P.C. (1997). Contribution of red blood cell aggregation to venous vascular resistance in skeletal muscle. *Am. J. Physiol.*, **272**, pp. H1020–H1032.

16. Cokelet, G.R., and Goldsmith, H.L. (1991). Decreased hydrodynamic restistance in the two-phase flow of blood through small vertical tubes at low flow rates. *Circulation Res.*, **64**, pp. 1–17.

17. Johnn, H., Phipps, C., Gascoyne, S., Hawkey, C., and Rampling, M.W. (1992). A comparison of the viscometric properties of the blood of a wide range of mammals. *Clin. Haemorheol.*, **12**, pp. 639–647.

Characterisation of Monocytes by Filtration of Undiluted Blood

Shelley-Ann Evans and Alison Cook

Cardiff School of Biosciences, Cardiff University, Wales

Abstract

The effect of increased adhesiveness and decreased deformability of leukocytes, following activation *in vivo*, can have a profound effect on flow through the microcirculation. Measurement of leukocyte deformability is therefore an important tool in the study of the pathology of vascular diseases. Although much work has been done on the rheological properties of lymphocytes and granulocytes, there is little information available on the larger mononuclear cells, the monocytes. To investigate monocyte rheology, blood obtained from 2 groups of healthy volunteers was filtered, undiluted, through 5 μm polycarbonate filters using a fully automated constant pressure filtrometer. Flow profiles were recorded over 300 seconds and the profiles were analysed by least squares fitting to an appropriate mathematical model.

In both young and elderly adults, the leukocytes fall into 4 distinct sub-populations in terms of their rheological properties. In both groups, the majority of leukocytes are in rapid equilibrium with the pores of the membrane, and are predominantly lymphocytes and granulocytes. The remaining leukocytes are not in equilibrium with the pores and analysis of the declining flow rate calculates both the number and flow properties of these cells. These cells equate numerically with the monocytes, and are sub-divided into 3 distinct sub-populations according to their rheological properties. Other workers have characterised monocytes into defined subsets on the basis of their size or phagocytic ability. Filtration of blood seems to be an acceptable way of measuring monocyte rheology, thusFlow profile of blood through 5νm filter. The points are experimental, and the line is the least squares fit avoiding purification which may alter their morphology, and the rheological heterogeneity of monocytes is confirmed.

1 Introduction

The properties of all constituents of blood, and any interactions between them, affect blood flow through the circulation. The properties of some constituents are more important in bulk flow through the large vessels, while the properties of other constituents are important only during flow through the microcirculation. In capillaries, the diameters of the blood cells are large when compared to the vessel diameter, and the deformability of individual cells, cell numbers and interactions between them become important in determining blood flow under these conditions. Leukocytes, although fewer in number than erythrocytes, offer a significant contribution to flow resistance in small diameter vessels, since their complex internal structure gives them a higher internal viscosity so their deformation is slower than that of erythrocytes [1]. In order to fulfil their physiological role of immune surveillance and response, leukocytes tend to roll or adhere to postcapillary vascular endothelium [2], and this interaction of leukocytes with the endothelium during immune and inflammatory events is dependent on a series of transient cellular adhesive events [3]. These events can result in decreased vessel lumen hence increased flow resistance and can temporarily block local capillary flow [4] and [5]. Furthermore, during prolonged low flow states polymorphonuclear granulocytes (PMN) and monocytes may be trapped in capillaries causing irreversible blockage and giving rise to capillary no-reflow phenomenon [6] and [7]. Activation of neutrophils and monocytes is accompanied by cytoskeletal organisation, which increases the internal viscosity of the cell and decreases the cell deformability.

The properties of PMNs and monocytes are extremely sensitive to environmental conditions and change on activation [8–10]. As mentioned earlier, impairment of leukocyte rheological properties, numbers and activation seem to be major factors involved in the onset and progression of ischaemic events. For these reasons, there has been a growing interest in measurement of leukocyte rheology. Filtration methods for measuring leukocyte deformability involve isolation and purification of cell sub-populations, in order to filter homogeneous or near-homogeneous populations of cells. The advantage of dealing with purified sub-populations of cells is that filtration profiles can be analysed with greater ease, since simpler mathematical models can be used for homogeneous suspensions. Conversely, purification of cells can be problematical for a number of reasons. Firstly, the isolation procedures used can be time consuming and possibly rheologically damaging, as they usually involve successive centrifugations on density gradients [11–13], and repeated washings in a variety of suspending media. As a result of this, there is the possibility that the purified cell suspensions may be metabolically or structurally altered by these isolation procedures. These problems are even more pronounced for monocytes, with usual isolation methods involving their adherence and subsequent detachment from plastic or glass petri-dishes [14].

Measuring the deformability of erythrocytes and leukocytes is therefore a complex business, and a variety of techniques have been used [15]. Filtration techniques are currently one of the most widely used methods for assessing leukocyte deformability, and involve measuring the flow of cell suspensions through a porous filter. The interpretation of the filtration properties of suspensions, in terms of recognisable physical parameters of the cells, requires mathematical models as a theoretical basis [16]. A pore transit time, the time taken for a single cell to completely traverse a pore, is commonly quoted. Quantifying pore blocking of the filter is possible from the decline in the initial flow rate, although some methods were designed to be insensitive to pore blocking [17]. To date, many investigators have used filtration procedures with little or no analytical techniques. However, precautions must be taken to avoid selection of an over-complex mathematical model [18] and [19] as more complex mixtures of cells are now being analysed by filtration techniques. Filtration analysis can therefore be problematical unless a rigorous and appropriate analytical approach is used.

Measurement of leukocyte deformability is therefore an important tool in the study of the pathology of vascular diseases. Although much work has been done on the rheological properties of lymphocytes and granulocytes, there is little information available on the larger mononuclear cells, the monocytes, which comprise about 5% of the total leukocyte count. This presentation documents a series of studies undertaken in order to evaluate the rheological properties of monocytes.

2 Methods

2.1 Blood samples and cell counting

Blood was collected from healthy volunteers from the antecubital vein, and collected into vacutainers (Becton Dickinson, Cowley, Oxford) containing tri-potassium EDTA. In all experiments, blood was used within 4 hours of venepuncture. For the monocyte purification techniques, blood was taken from healthy laboratory personnel. For the filtration studies, two groups of volunteers were used, with all people having no history of haematological or vascular disorders. Group A were 4 females (age 21–25 years); Group B were 12 males (aged 53–79 years). All cell counts were performed on a Serono-Baker System 9000 automated cell counter. This instrument counts lymphocytes, granulocytes and "mid range cells" which, in comparison with a Technicon H^*1 instrument, are comprised about 80% monocytes. The number of mid range cells is therefore taken as an approximation of the monocyte count.

2.2 Monocyte purification

Monocytes were purified by 2 different methods. The first was a one-step, density gradient centrifugation using J.N.Prep 1.068 (TechGen International Limited, London). Leukocyte rich plasma was prepared from blood by sedimentation as previously described [20], and 10 ml was gently layered onto 5 ml of the gradient. This was centrifuged at 600 g for 15 minutes, after which the appropriate band was removed and the cells were washed sequentially in phosphate buffered saline (PBS) and PBS with 1.5 mg/ml EDTA and 5% plasma. The second method used for monocyte purification was an adherence method similar to that described by Nash et al. [12]. Briefly, mononuclear cells were prepared by density gradient centrifugation, and were incubated in RPMI cell culture medium in plastic petri dishes for 60 minutes. Non-adherent cells were removed by aspiration, and the adherent monocytes were detached using a solution of PBS containing trypsin

(0.05%) and EDTA (0.02%). The cells were then washed as described above for mononuclear cells.

2.3 Electron microscopy

Adherent cells, detached from petri dishes (as above) were fixed as described elsewhere [21]. The microscopy was performed using a Phillips EM 400T with scanning attachment.

2.4 Filtration

Nuclepore polycarbonate filters (5 μm pores) were mounted in a fully automated and custom built filtrometer described elsewhere [22]. Each membrane was calibrated by measuring the flow rate of suspending medium (plasma) before the filtration of blood. The instrument uses a strain gauge transducer and weights are gathered at 20 millisecond intervals by a computer. There is no turbulent flow nor lag period in this filtrometer. Flow profiles of volume against time were recorded over 300 seconds at a pressure of 711 Pascals.

2.5 Theoretical considerations

The earliest measurable flow rate reflects the flow properties of plasma, red blood cells, lymphocytes and granulocytes, that is, they are in rapid equilibrium with the pores of the filter. The initial flow rate is reduced by a population of slow leukocytes representing 5% of the total leukocyte count, and all discussion from herein concerns these cells only, termed "slow leukocytes" or "SL". The possible presence of either one or two flowing populations of cells, plus one pore blocking population of cells (PB) was investigated; pore blocking particles include any slow flowing cells which have not been given sufficient time to pass through the pores in the filter. As the flow rate continuously declines, calculation of the numbers and flow properties of these slow flowing leukocytes is possible. A brief overview of the mathematical models used is given below; but full descriptions are published elsewhere [23].

The flow rate of blood through the filter is given by:

$$\frac{dV}{dt} = Vi(T - P_{SL1} - P_{SL2} - P_{PB})$$

and

$$Vi = V_b\omega.$$

Where Vi is the initial flow rate, V_b is the combined flow rate of plasma and red cells, and ω is the proportion of pores in the filter not occupied by lymphocytes and granulocytes (which are in rapid equilibrium with the pores); T is the total number of pores in the filter and P_{SL1}, P_{SL2} and P_{PB} are the number of pores occupied by each of the 3 assumed populations of slow leukocytes. The rate of occupation of pores by 2 populations of flowing cells and a pore blocker is described by:

$$\frac{dP_{SL1}}{dt} = k_5\omega(T - P_{SL1} - P_{SL2} - P_{PB}) - k_6P_{SL1}$$

$$\frac{dP_{SL2}}{dt} = k_7\omega(T - P_{SL1} - P_{SL2} - P_{PB}) - k_8P_{SL2}$$

$$\frac{dP_{PB}}{dt} = k_9\omega(T - P_{SL1} - P_{SL2} - P_{PB})$$

Where: $k_5 = [SL_1].V_b$; $k_7 = [SL_2].V_b$; $k_9 = [PB].Vb$; k_6 and k_8 are the proportion of pores, occupied by SL_1 or SL_2 respectively, being evacuated in unit time. Pore transit times of the kinetic populations are calculated as reciprocals of these rate constants k_6 and k_8 respectively.

The above equations are integrated to construct theoretical curves using routines from the NAG library, and fitted to the experimental data. The "goodness of fit" between the experimental and theoretical curves was checked using the F-test. This checks whether an improvement in the sums of squares of deviations of experimental from theoretical values achieved by a more complex model, is more than expected from the presence of extra constants in the equation. The analysis gives the best values for Vi and the rate constants.

3 Results

3.1 Monocyte purification

Purification of monocytes using a commercially available, one step density gradient was neither reproducible nor efficient. For 10 separate volunteers, the monocyte yield ranged from 2–36%, with purity ranging from 3-45%. Thus, it was difficult to get enough cell suspension for rheological studies. The monocytes purified by the adherence and detachment method were examined by electron microscopy which showed that the cells had grossly altered morphology, seen in Figure 1 below. In mononuclear suspensions (Figures 1(a) and 1(b)) the cells appear rounded, with their outer membranes having prominent ridges and folds. However, after undergoing adherence and detachment, the monocytes are flattened in appearance with a dramatically altered surface morphology, as seen in Figures 1(c) and 1(d). Further rheological investigation of these cells was deemed to be undesirable for 2 reasons. Firstly, the dramatic alterations seen in surface morphology would certainly adversely affect any rheological investigations. Secondly, complete recovery of all monocytes was impossible to achieve without some loss in cell number, so it was difficult to obtain monocytes in sufficient numbers for rheological studies. Rheological investigations were therefore not undertaken on either of these isolated populations of monocytes.

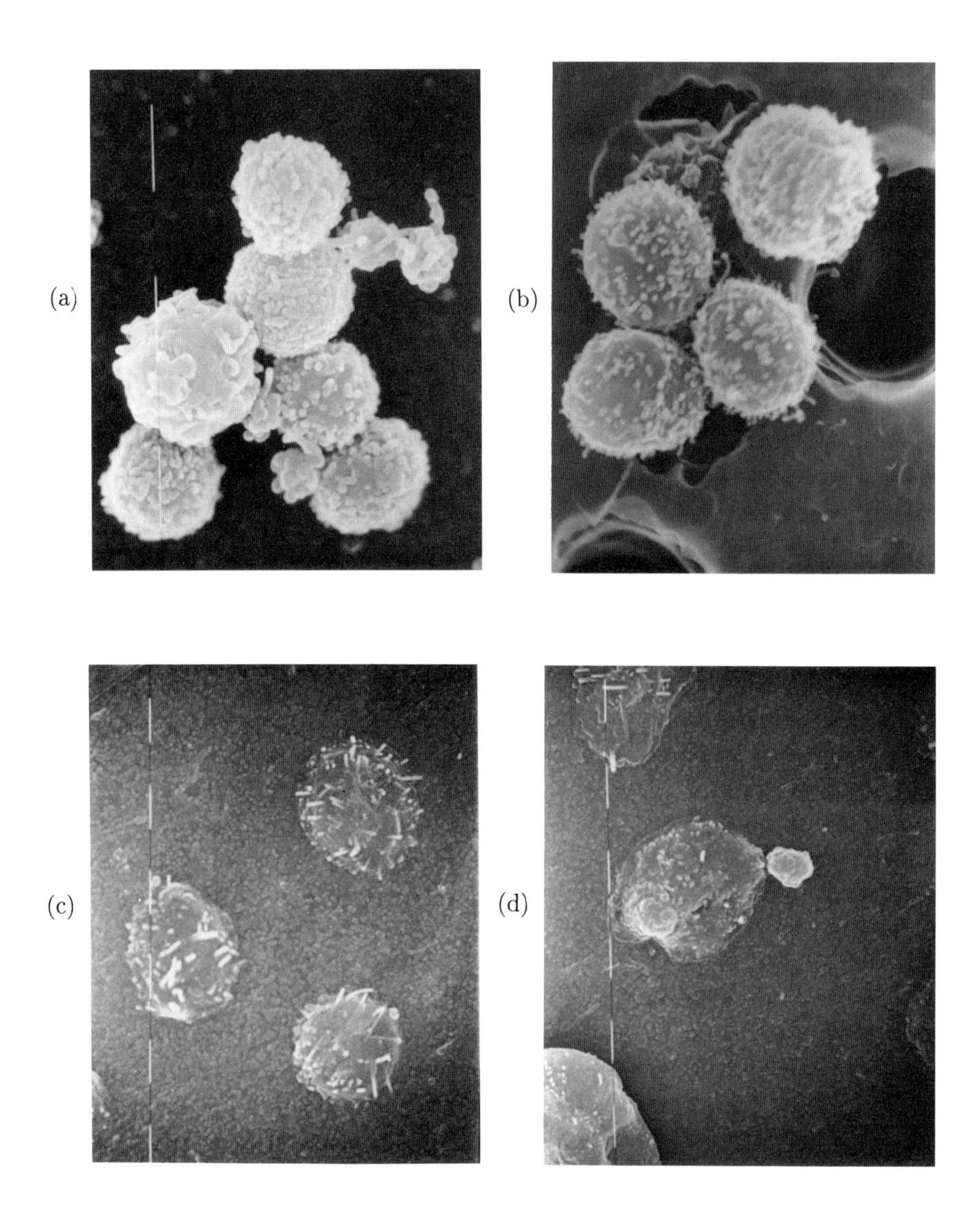

Figure 1. Electron micrographs of monocytes prior to and after adherence and detachment. Mononuclear cells are shown in (a) and (b); whilst monocytes (after undergoing adherence and detachment) are shown in (c) and (d). The scale bars shown are $5\mu m$

3.2 Filtration studies

For the filtration studies, performed with undiluted blood, the haematological data is shown in Table 1. Apart from the expected lower red cell count in Group A (young females), there were no significant differences in the other measured parameters.

A typical flow profile of blood is shown in Figure 2 below. Analysis of these curves, as described briefly above, yields pore transit times and concentration of all particles present, and the data is shown below in Tables 2 and 3. The more complex model of 3 particles (SL_1, SL_2 and PB) invariably gives a better fit than 2 particles (SL_1 and PB), with the improvement in fit, as assessed by the F-test, being significant in 92% of cases.

In both groups, the monocytes (about 5% of the total leukocyte count) fall into 3 rheologically distinct populations. The transit times are the same for both groups, and the third population of slow leukocytes was recorded as pore blockers in these experiments.

The concentrations of cells in these 3 rheologically distinct groups is shown in Table 3. The number of mid range cells may be a little higher in Group B (p = 0.08) and this is reflected in a marginally higher concentration of

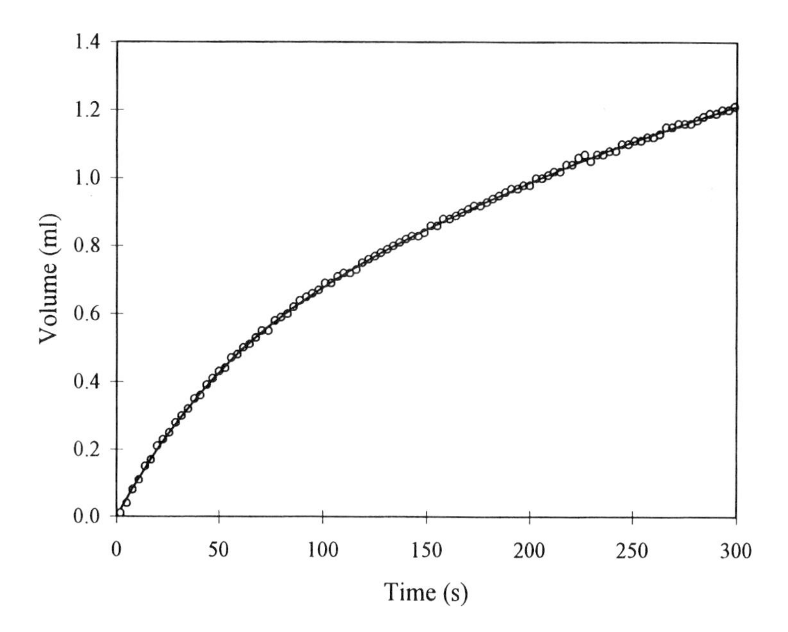

Figure 2. Flow profile of blood through 5μm filter. The points are experimental, and the line is the least squares fit

Table 1. Haematological data of blood from Groups A and B

	Group A mean ± SEM	Group B mean ± SEM
Red blood cells ($10^9/ml$)	4.46 ± 0.21	5.08 ± 0.10
Total leukocytes ($10^6/ml$)	6.25 ± 0.43	6.91 ± 0.54
Mid range cells ($10^6/ml$)	0.37 ± 0.03	0.48 ± 0.05

Table 2. Pore transit times (in seconds) for Groups A and B

Cell Population	Group A mean ± SEM	Group B mean ± SEM	p Value (Unpaired t Test)
SL_1	34.8 ± 1.4	31.7 ± 3.5	0.43
SL_2	147.5 ± 2.5	145.8 ± 11.9	0.89
PB	>300	>300	0.97

Table 3. Concentrations of particles kinetically counted and compared to mid range count

Cell Population	Group A mean ± SEM	Group B mean ± SEM	p Value (Unpaired t Test)
SL_1 ($\times 10^6/ml$)	0.12 ± 0.02	0.19 ± 0.03	0.04
SL_2 ($\times 10^6/ml$)	0.11 ± 0.02	0.10 ± 0.01	0.73
PB ($\times 10^6/ml$)	0.09 ± 0.02	0.09 ± 0.01	0.97
$SL_1 + SL_2 + PB$ ($\times 10^6/ml$)	0.32 ± 0.06	0.38 ± 0.03	0.39
Mid range count ($\times 10^6/ml$)	0.37 ± 0.03	0.48 ± 0.05	0.08

SL_1 ($p = 0.04$) but all the differences are small. The total concentration of all slower leukocytes (SL_1, SL_2 and PB) is marginally less than the mid range count but the difference does not reach statistical significance.

4 Discussion

The ideal method for isolation of monocytes should be simple, require only standard laboratory equipment, produce pure cells at maximum yield, and not alter the morphology or cytochemistry of the cells in any way. Such a method does not exist. Leukocytes, particularly PMNs, are very sensitive to changes in environment which can trigger their activation, thus dramatically altering their rheological properties. These problems are even more pronounced for monocytes, with usual isolation methods involving their adherence and subsequent detachment from plastic or glass petri-dishes [14], although historically, purification techniques based on adherence have not been extensively employed for preparation of monocytes if the cells are required to be in suspension. Since their morphology has been shown here to be altered by such procedures, their rheological properties may also differ. Also, the purified cells are not studied in their native plasma environment. Analysis of filtration profiles of blood offers the possibility of determining the properties of sub-populations of cells with minimal manipulation, thus avoiding possible detrimental effects.

With regards to the filtration studies, the slow leukocytes represent 5% of the total leukocytes which equates with the percentage of monocytes present. The monocytes fall into 3 rheologically distinct sub-populations of cells. Other workers have characterised monocytes into defined subsets on the basis of size [24], phagocytic ability [25] and modal volume [26] with all workers agreeing on the existence of at least 2 major sub-populations. The rheological data here leads to similar conclusions, in that monocytes are a heterogeneous cell population. The concentration of cells in these subsets of slow leukocytes are likely to vary in different patient groups, particularly in those with severe circulatory problems whose phagocytic cells may be "primed". This may lead to cells being displaced into the slow leukocyte sub-populations, hence shown by an increase in the number of SL_1, SL_2 or pore blockers.

Bibliography

1. LaCelle, P.L. (1986). Alterations by leukocytes of erythrocyte flow in microchannels. *Blood Cells*, **12**, pp. 179–189.

2. House, S.D., and Lipowsky, H.H. (1988). In vivo determination of the force of leukocyte-endothelial adhesion in the mesenteric microvasculature of the cat. *Circ. Res.*, **63**, pp. 658–668.

3. Springer, T.A. (1990). Adhesion receptors of the immune system. *Nature*, **346**, pp. 425–434.

4. Bagge, U., and Branemark, P.I. (1977). White blood cell rheology. An intravital study in man. *Adv. Microcirc.*, **7**, pp. 1–17.

5. Bagge, U., Amunson, B., and Lauritzen, C. (1980). White blood cell deformability and plugging of skeletal muscle capillaries in haemorrhagic shock. *Acta Physiol. Scand.*, **108**, pp. 159–163.

6. Barroso-Aranda, J., Schmid-Schonbein, G.W., Zweifach, B.W., and Engler, R.L. (1988). Granulocytes and no-reflow phenomenon in irreversible haemorrhagic shock. *Circ. Res.*, **63**, pp. 437–447.

7. Engler, R.L., Schmid-Schonbein, G.W., and Pavelec, R.S. (1983). Leukocyte capillary plugging in myocardial ischaemia and reperfusion in the dog. *Am. J. Pathol.*, **111**, pp. 98–111.

8. Nash, G.B., and Meiselman, H.J. (1986). Rheological properties of individual polymorphonuclear granulocytes and lymphocytes. *Clin. Hemorheol.*, **6**, pp. 87–97.

9. Nash, G.B., Jones, J.G., Mikita, J., Christopher, B., and Dormandy, J.A. (1988). Effects of preparative procedures and cell activation on flow of white cells through micropore filters. *Brit. J. Haem.*, **70**, pp. 171–176.

10. Pecsvarady, Z., Fisher, T.C., Fabok, A., Coates, T.D., and Meiselman, H.J. (1992). Kinetics of granulocyte deformability following exposure to chemotactic stimuli. *Blood Cells*, **18**, pp. 333–352.

11. Lennie, S.E., Lowe, G.D.O., Barbenel, J.C., Forbes, C.D., and Foulds, W.S. (1987). Filterability of white cell subpopulations, separated by an improved method. *Clin. Hemorheol.*, **7**, pp. 811–816.

12. Nash, G.B., Jones, J.G., Mikita, J., and Dormandy, J.A. (1988). Methods and theory for analysis of flow of white cell subpopulations through micropore filters. *Brit. J. Haem.*, **70**, pp. 165–170.

13. Schmaltzer, E.A., and Chien, S. (1984). Filterability of subpopulations of leukocytes: effect of pentoxifylline. *Blood*, **64**, pp. 542–546.

14. Fischer, D., and Koren, H. (1981). Isolation of human monocytes. *Methods for Studying Mononuclear Phagocytes*, Academic Press, pp. 43–47.

15. Matrai, A., Whittington, R., and Skalak, R. (1987). Biophysics. *Clinical Hemorheology*, Martinus Nijhoff, Dordrecht, pp. 9–71.

16. Jones, J.G., Evans, S-A., and Adams, R.A. (1994). Bulk flow through micropore membranes for analysing blood cell rheology in clinical research. *Clin. Hemorheol.*, **14**, pp. 149–169.

17. Hanss, M. (1983). Erythrocyte filtrability measurement by the initial flow rate method. *Biorheology*, **20**, pp. 199–211.

18. Kooshesh, F. (1989). Measurement of the deformability of red blood cells. *Ph.D. Thesis*, University of Wales.

19. Evans, S-A. (1990). Studies on the deformability of red and white blood cells. *Ph.D. Thesis*, University of Wales.

20. Mikita, J., Nash, G.B., and Dormandy, J.A. (1986). A simple method for preparing white blood cells for filterability testing. *Clin. Hemorheol.*, **6**, pp. 635–639.

21. Jones, J.G., Holland, B.M., Humphrys, J., and Wardrop, C.A. (1985). The flow of blood cell suspensions through 3 μm and 5 μm nuclepore membranes; a comparison of kinetic analysis with scanning electron microscopic examinations. *Brit. J. Haem.*, **59**, pp. 541–546.

22. Adams, R.A., Evans, S-A., and Jones, J.G. (1994). Characterisation of leukocytes by filtration of diluted blood. *Biorheology*, **31**, pp. 603–615.

23. Cook, A.M., Evans, S-A., and Jones, J.G. (1998). The filterability of leukocytes in undiluted blood. *Biorheology*, **35**, pp. 119–130.

24. Dougherty, G.J., Dougherty, S.T., and McBride, W.H. (1989). Monocyte heterogeneity in human monocytes. *Human Monocytes*, Academic Press, pp. 71–78.

25. Chiu, K.M., McPherson, L.H., Harris, J.E., and Braun, D.P. (1984). The separation of cytotoxic human blood monocytes into high, low phagocytic subsets by centrifugal elutriation. *J. Leukocyte Biol.*, **36**, pp. 729–737.

26. Arenson, E.B., Epstien, M.B., and Seeger, R.C. (1980). Volumetric, functional heterogeneity of human monocytes. *J. Clin. Invest.*, **65**, pp. 613–618.